THE CREATION OF MATTER

THE CREATION OF MATTER

THE UNIVERSE FROM BEGINNING TO END

HARALD FRITZSCH

Translated by Jean Steinberg

Basic Books, Inc., Publishers   New York

Library of Congress Cataloging in Publication Data

Fritzsch, Harald
  The creation of matter.

  Translation of Vom Urknall zum Zerfall.
  Includes index.
  1. Cosmology. 2. Big bang theory. 3. Matter—
Constitution. I. Title.
QB981.F7413 1984   523.1   83-46089
ISBN 0–465–01446–1

English Translation
Copyright © 1984 by Basic Books, Inc.
Originally Published in German as
Vom Urknall zum Zerfall: Die Welt zwischen Anfang und Ende
© 1983 by R. Piper & Co. Verlag, Munich
Printed in the United States of America
Designed by Vincent Torre
10   9   8   7   6   5   4   3   2   1

To Oliver and Patrick

Contents

Contents

Preface

THIS BOOK about the origin and development of the universe is meant not only as a source of information but also as a guidepost in an age in which all of us are saturated with information and in which it has become difficult to find one's way. For that reason I have confined myself to the most important aspects of physics and cosmology. The major problem in writing a book such as this concerns sifting out what to include from what to leave out. Thus I have refrained from discussing the theory of relativity in greater detail not because I do not consider it important but because it is possible to understand a great many aspects of cosmology, such as the origin of matter soon after the Big Bang, without a detailed knowledge of the theory of relativity. On the other hand, I have discussed the significance of quantum mechanics for the structure of matter in greater detail because I feel that one cannot understand the modern theory of matter without some understanding of quantum phenomena.

To understand what took place at the primary explosion about 20 billion years ago and why the world turned out to be as we find it today, one has to know something about the structure of the cosmos and the structure of matter. A substantial portion of this book is devoted to the transmittal of that knowledge. Chapters 9 through 13 deal with the actual theme of this work, with the Big Bang. The book closes with a discussion of the remote future. Problems of philosophy and religion are considered in the three final chapters.

I have discussed many of the problems treated here with my co-workers at the University of Munich, at the Max Planck Institute for Physics at Munich, and especially at the European Nuclear Research Center, CERN, at Geneva, where most of this book was written. I have profited greatly from these discussions, and I am grateful to all participants in them. I would also like to thank Martin Kessler and his collaborators at Basic Books for their efforts in producing this book, and Jean Steinberg for her excellent translation of the German text.

HARALD FRITZSCH
Munich, Summer 1984

THE CREATION OF MATTER

Prologue

I N the beginning there was nothing, neither time nor space, neither stars nor planets, neither rocks nor plants, neither animals nor human beings. Everything came out of the void. It all began with space and time and a very hot plasma composed of quarks, electrons, and other particles. This plasma cooled off rapidly; protons, neutrons, atomic particles, atoms, stars, galaxies, and planets formed. Finally life sprang up in many solar systems of the universe—in one case, on a planet of a most ordinary star situated on a spiral arm of a galaxy at the rim of a large cluster of galaxies. There, in the course of 4 billion years, plants and animals, and eventually human beings, developed out of the simplest organisms.

Originally human beings thought that they stood in the center of the universe, that the world was made for them alone. They invented gods and divine schemes that they believed governed the world. The world of human beings was small. The firmament enveloped the world like a protective skin.

Five hundred years ago, about 20 billion years after the birth of the universe, after the Big Bang, we human beings began a systematic exploration of our environment and ourselves. Toward the end of the millennium we began to see that the diversity of the world can indeed be explained. All matter in the universe, including ourselves, is composed of two types of minute building blocks: quarks and the particles of the atomic shell, electrons. We have come to understand that we are not the center of the universe, that we merely live inside a rather unremarkable galaxy. We still have not made contact with other inhabitants of the universe in other solar systems, but we sense that we may not be alone. We have also learned that we are the product of a complex yet rationally comprehensible process of development, determined by both history and the interplay of chance and necessity. We have come to realize that we must live without gods, that we alone are responsible for our fate. We have begun to sense that the universe does not hold the answer to questions about the meaning of life, that it is up to us to find the answers.

The end of the twentieth century is ushering in what may well be the most important period of human civilization. We have begun to realize that the meaning of our existence lies *in* our very existence and in our constant search for the answers to questions for which no universally valid answers exist.

Our civilization dates back tens of thousands of years, but only comparatively recently, within the last few hundred years, have scientists, along with engineers and technicians, embarked on a systematic study of nature and the natural processes that occur in it. The application of their findings has brought about profound changes in all our lives.

Modern physics, astrophysics, chemistry, and biology offer us deep insights into the structure of the universe. In recent years physicists, and astrophysicists in particular, with the help of gigantic accelerators and telescopes, have gained understanding of the processes that took place some 20 billion years ago, at the birth of our cosmos. Yet only a few among them have realized that something remarkable is taking place today, namely that we are in the process of developing a unified picture of the entire universe. We now know far more about the development of the universe, including the origin of life, than would have been thought possible only yesterday.

There is no denying that the great strides in science and technology have brought unprecedented material well-being to many. Yet the other side of this development is equally obvious. Established traditions and ties, values and norms, are being questioned. A sense of satiety, of senselessness and insecurity, even of fear, is spreading. Many among us, particularly the young, with only a scanty knowledge of science, have lost faith in the future and feel that they have been misled. And the schools, where the sciences often are poorly taught, contribute their part to this state of affairs.

One of the reasons for this trend undoubtedly is connected with the state of contemporary science and technology, both of which are split up into so many areas of specialization that no scientist can claim to possess a more or less comprehensive overview. Thus no physicist today can take in the entire realm of physics. We have atomic physicists, nuclear physicists, molecular physicists, solid-state and particle physicists, astrophysicists, and so forth. This specialization often gives rise to the false conclusion that it has become impossible to gain a coherent view of the science of physics as a whole. The world of modern science, one frequently hears, is so complex that the design of a comprehensive theory of the world appears to be a hopeless quest. We shall have to make do with partial views. I for one, however, believe the exact opposite to be true. New insights into the structure of matter and the universe gained by elementary-particle physics and astrophysics in particular indicate that the structure of the cosmos is in fact quite simple. A "comprehensive view" of the universe lies within the realm of the possible.

Five hundred years ago a map of the world showing all the oceans and continents was heralded as a major accomplishment. Now, thanks to modern science, we can draw a "map" of the entire universe. Like the first maps of the world in the days of Columbus, this one is, of course, very rough and not without flaws. Continents may still be missing. But the fact that we are able to draw such a map at all is remarkable and is certain to bear on the future. Today the entire cosmos lies before us. The fogs that enshrouded it are lifting. Like hikers climbing a mountain early in the morning, we now for the first time are able to see exactly where we stand. Where are we, and why? Are we on a journey from nowhere to nowhere or does the universe hold out a meaning?

One thing, however, is certain: even if we succeed in solving all the riddles of nature we will not have touched on the human problems. Who sets the values without which life would have no meaning? Does religion still make sense in the face of a science which is able to explain the Creation without divine intercession?

One of the most significant contributions of modern science is the realization that each one of us, enmeshed in the web of natural processes, matters. We are a part of a whole, not separate from the rest of the universe. At the same time we are the product of a long history—and also the makers of history. It is my belief that this awareness should fill us not only with modesty but also with pride and self-understanding, and that, armed with this awareness, we will better be able to develop the values essential to our society.

CHAPTER 1

The Dance with the Ocean

> Youngsters, tired of badly programmed computers and people who act like badly programmed computers, are turning to tarot cards and charlatans.
> —MURRAY GELL-MANN

T H E idea for writing this book came to me on the beach of Santa Barbara, California, in May 1981. I was attending a meeting of scientists from all corners of the globe, mostly elementary-particle physicists and astrophysicists, at the Santa Barbara campus of the University of California. We were engaged in a discussion about the events and natural processes that took place shortly after the birth of our cosmos and that continue to play a significant role in the structure of the universe and its matter—including the earth and ourselves. Basically the conference was concerned with the unity of physics, the idea of the unification of all forces and its implications for the cosmos.

The Dance with the Ocean

The Santa Barbara campus must be one of the most beautiful places in the world. Looking out on the rugged mountains where condors still nest, it stretches along the sun-drenched shores of the Pacific. On a clear day one can see the islands that dot the shore. Sea lions and dolphins cavort in the sea, and gulls, sandpipers, and pelicans populate the beach.

Between sessions I would walk on the beach, and there one day I met the young student who became the catalyst for this book. The first time I saw her she was dancing on the beach, eyes closed and face turned toward the sun, keeping time with the roll of the surf. I stopped to watch. She did not appear to notice me, and so I sat down in the sand to watch this somewhat unusual performance. I was fascinated by this dance in harmony with the play of the waves and the sun, by the utter self-absorption of the dancer. After a while dancer, ocean, and birds became part of a whole—different aspects of the same thing. At that moment this young woman obviously was in complete harmony with herself and with nature.

Yet my presence did not go unnoticed. After a time I found myself looking into a smiling face flushed by wind and water. "I was quite taken by your dance," I said, trying to hide my embarrassment.

"Thank you," she answered. "I hadn't really counted on an audience. I often come here, usually after class. I think I might not make it here if it were not for the ocean and the beach."

We began to talk. As I had suspected, she was a student, a sophomore majoring in philosophy.

"You know, after class I feel as though something is missing. In the lectures the world is served to us chopped up into bits. The larger view is lost. Here on the beach it's altogether different. When I dance in the wind I know that I am part of it, of the waves, the ocean, the sea lions, the gulls. Here I feel like part of the world—the world is me, the sea lions, and everything else."

We stayed on the beach for quite some time; I told her what had brought me to Santa Barbara. We spoke about the search of modern science for a rational explanation for all natural phenomena. It was not easy to make her understand the problems that modern science sought to solve. She refused to recognize that one could arrive at truth through research and experimentation, through "dissection," as she called it.

"You see," she finally said, "when I'm here at the beach, I feel that I'm part of nature, at one with the sea and the birds above me. That is my truth. That's all I need. You physicists and chemists and biologists always pick everything apart. You examine the waters of the ocean, you dissect the fish and sea lions and birds. You destroy everything that gives me pleasure. I don't need your truths. All I need is my own experience. I hope I haven't hurt your feelings by saying this. But now I must go. See you later." And she rushed off, back to the campus.

The following day I met her again, this time not on the beach but in the school cafeteria. We had a long talk about a number of things, including physics.

"And so you are interested in the puny truths of physics after all," I said somewhat ironically.

"I must have been pretty obnoxious down on the beach," she said, laughing.

"At least you seemed very certain of your philosophy of the great, whole truth."

"I am," she said, "but only when I'm at the beach. Up here it's different. Sometimes I think that I'm two different people. That confuses me, makes me unsure. Perhaps I'm wrong and my 'beach philosophy' is nothing but an illusion."

"I think that you're basically right. Both of us are. Neither of us can do without that which the other one sees. All of us are constantly subjected to a stream of impressions which we must absorb and which often challenge us to act. The totality of these impressions may be said to be the entirety of the world of which you spoke. I don't deny that this entirety, this constant interaction of all objects, exists. But we cannot live solely with this entirety. Our civilized world came into being because we learned to divide the world into separate areas, to classify and dissect it. I admit that some scientists see the world only in terms of the details they discovered in their dissections. Thus a physicist might see a bird as nothing more than a collection of atomic nuclei and electrons able to hover in the air. No doubt those who see things this way lose a lot. I see the scientific method of dividing the world into many individual aspects merely as a tool in gaining a better understanding of the totality of the cosmos—of its complex interconnectedness, its individual aspects and wholeness. Niels Bohr, one of the founders of atomic theory, used to say: 'We are at the same time spectators and actors in the great drama of nature.' Wholeness and detail—these are the two extremes in the spectrum of possibilities that we in the sciences are investigating. When you say that you can do without the detail and will accept only the whole, you are foregoing an important part of all the possible ways of looking at nature. Moreover, such a position cannot really be strictly maintained."

The young woman looked at me quizzically. "But who

says that I can't forego all these possible ways? They don't interest me."

"Experience shows," I answered, "that with nothing but your whole you could not exist here at all. If we were not to split up the world into individual details we could not survive in the long run. We would have to give up not only science but all technology. Our modern civilization with all its merits and faults would collapse, and we would have to return to the forests of our forebears. I don't think that you're one of those extremists who really would want that to happen. Science is a systematic method for the investigation of the connections between detectable phenomena. However, that does not mean that scientists split up the world into little pieces at every available opportunity. When I watched you dance down at the beach yesterday I was fascinated, and nothing was further from my mind than to dissect the scene into its components. I will remember it in its totality—the surf and your dance in the wind. The rational analysis of the world by science and the emotional comprehension of the world as an entity—these contrasts belong together like hot and cold. Our world is too complex to be viewed through only a single aspect, whether physical, chemical, or biological. Many different views are needed for a more or less complete understanding. Let me tell you a little story.

One day Werner Heisenberg, one of the founders of quantum theory, and his disciple Felix Bloch were out walking. Bloch used this opportunity to bring some new mathematical ideas about the structure of space to Heisenberg's attention. Heisenberg did not seem entirely convinced of the importance of Bloch's exposition. At any rate, he interrupted him, saying, 'Space is blue and birds fly through it.' With that he was throwing an entirely different aspect of 'space' into the discussion, one that seemed important to him at that moment."

The Role of Science

My conversations with the young woman stayed in my mind for some time. Here we were, physicists and astrophysicists, preoccupied with the natural universe and its history, while at the same time students at this very university were rejecting the scientific approach to the study of the world, were even turning this rejection into a symbol. Wouldn't it be interesting, I asked myself, to delve more deeply into this contradiction? Is the chasm dividing the rational view of the world and the irrational totality of the cosmos indeed unbridgeable?

The scientific, rational method of looking at the world seeks to establish relationships between different objects and processes, to find causal connections. Therein lie both the strengths and limitations of scientific knowledge. It cannot define values and goals, not even the goals of science. Even if we succeeded in unraveling all the riddles of nature and solving all its problems, we would not have touched upon fundamental human questions about the meaning and goals of life. Albert Einstein once said, "It is ... clear that knowledge of what *is* does not open the door directly to what *should be.*"* A rational grasp of the world is quite possible but it is not enough.

Even if rational knowledge cannot give us values and

*Albert Einstein, *Out of My Later Years* (Westport, Conn.: Greenwood Press, 1970), p. 22.

13

goals, it can do something else. It can tell us where we stand in the universe. I believe that this factor plays a crucial role in the establishment of individual values and goals, as well as the values and goals of society as a whole.

The lay public tends to have an erroneous picture of the work and conceptions of scientists. The experience of nature as a whole and the investigation of nature, the establishing of natural laws through research, are not two diametrically opposed activities but merely two different aspects of one and the same thing. Most true scientists do not work only in order to develop new techniques that may at some time or other serve (or fail to serve) mankind, but because they are driven to understand why processes of nature follow one particular observed pattern instead of another. Their conceptualizations make it possible to predict the future course of a given natural process. And seeing one's prediction come true, one's intellectual curiosity satisfied, is the ultimate hope and reward of every scientist.

Skepticism toward science and technology is the order of the day in almost every developed country. Many people are interested in astrology and the occult, in things that have nothing in common with a rational understanding of reality.

To some degree this state of affairs is understandable. What, after all, does the general public see when looking at the sciences and technology? They see a society that treats the possibilities science has opened up to us with casual disdain. They see waste of raw materials and energy, they see cities being turned into concrete wastelands, they see superhighways despoiling the landscape.

In the past we cherished the illusion that everything was quantifiable and that everything that stood in the way of quantification could be ignored. But our lives are filled with phenomena that cannot be expressed in figures —abstractions like freedom or individual self-awareness, or the awe inspired by towering mountains.

Considering the fact that we possess a faculty that we call common sense, it is astonishing how often the decisions of governments and mass organizations, of political parties and big business, violate that common sense. Decisions are made on the basis of a wealth of facts and quantitative data compiled by special-interest lobbies, influenced by the demands of trade unions and the requirements of party politics.

In small enterprises good executives proceed differently. Faced with important decisions, such as substantial expenditures, they study all available data and work out alternate strategies and procedures. Having done so, they do not base their decisions on a specific quantifiable factor but, relying on intuition, also take imponderables into account. They sense which course is the right one for them.

It seems to me that the giant organizations of modern society have lost this instinct for what is the right course. The chief executive of a large company, who day after day is faced with crucial decisions, once told me, "So long as you have to make decisions whose consequences are obvious, your job isn't important. You become important only when you begin to realize that you don't understand the things you are dealing with." I think that politicians, government officials, and top managers will agree if I say that in this sense theirs are important positions.

Scientists frequently find themselves in the same position as business executives. If I, for example, want to examine a new scientific theory, I begin by studying the factors leading up to that theory. Then I ask myself, Is this theory on the right track? Does it make sense to pursue it any further? I invariably answer these questions intuitively, without reference to quantifiable factors. This method has saved me a lot of time, for it has enabled me to decide relatively quickly whether or not to pursue a new hypothesis. As in everyday life, in science too it is often a matter of what one does not do rather than what one does. Intuition is a valuable asset in separating the wheat from the chaff.

When Einstein finished working on his theory of general relativity he was fully convinced that it precisely described the phenomenon of gravity. Some years later his predictions regarding the deflection of light by the gravitational field of the sun proved to be correct. "What would you do if your prediction had not come true?" a reporter asked Einstein. His answer was not altogether modest: "I would feel sorry for God had He missed this opportunity." Einstein knew intuitively that he was right.

A New World View

Another reason why many people, particularly among the young, are not interested in the sciences and technology is, I believe, connected with the present organization of these fields. Our universities and colleges, and the lower-level schools in particular, have for decades shown a fatal trend toward dividing the world into different disciplines —a division which in fact does not exist. As a result, we have many highly qualified specialists, particularly in the exact sciences and technology, who know everything about their specialty, but know very little overall. Victor Weisskopf at M.I.T. has put this very succinctly: "An expert is someone who knows more and more about less and less, until finally he knows everything about nothing."

In addition, there has developed a tendency, especially among scientists, toward the glorification of expertise and a distrust of those who dare to look out beyond their own specialty and voice opinions on problems in other areas. A friend of mine, the physicist and Nobel laureate Murray Gell-Mann, has commented on these attitudes:

> Let me mention too the widespread failure to explain the overwhelming relevance of learning, of understanding, of analyzing, of using reason to approach the world and its problems, and the absolute necessity of using science and technology no matter what we want to do with our technically complex world, even if we want to make it less complex. "Learn this because I tell you to learn it, memorize pages 23 to 54," so often is what we say, instead of explaining how learning helps us to be complete and effective human beings.
>
> I believe that narrow rationality, pervading government, universities, industries, and other parts of our national and even international life, is provoking a wave of insufficient rationality. Youngsters tired of the tyranny of badly programmed computers, and of people who act like badly programmed computers, are turning to tarot cards and charlatans. . . .
>
> In an age of great technical complexity and of impressive scientific advances, we scientists and teachers have largely failed to communicate to the public, to students not specializing in science, and even to students in technical disciplines outside of our own, the meaning and importance, let alone the beauty and excitement of science. Do we expend a great deal of effort in trying to raise the standards of science and technology journalism? When we teach science courses for the arts or humanities students do we communicate successfully the world view that is emerging in molecular biology, or in geophysics, or in

particle physics, or in astrophysics? Do we explain what real research is like, and introduce the student to life and work in the laboratories nearby? Do we convey to the students the dialectic by which scientific discoveries are made and recognized? Do we explain the relevance of scientific and technical developments to political decisions and to the life of every human being on the planet? Or do we rather have the students memorize a few of the laws from an elementary physics book and regurgitate them on an examination?*

Most people, on hearing the words *science* and *technology,* think of more or less boring lectures in school. We still tend to think of *physics* as a science dealing with falling or rolling balls, with electricity and steam engines. Almost no one in school learns that modern physics—be it quantum mechanics, elementary-particle physics, astrophysics, or biophysics—is in the process of transmitting to us a new view of the world, and that the survival of life on this planet probably depends on an understanding of this world view. The bold idea of modern times since Newton lay in the vision that we could learn to understand the cosmos through thought and the active observation and exploration of nature. This assumption has proved correct, and the past fifty years have seen a significant evolution of thought. Today, many problems that once were considered the domain of religion or philosophy are being answered, or at least investigated, by modern science.

Modern physics examines the structure of matter at minute distances, which are more than a billion times as small as the diameter of an atom. Astronomers and astrophysicists penetrate more and more deeply into the universe and may perhaps have advanced to its limits, to the

*Murray Gell-Mann, *Physics Today,* May 1971.

boundaries of space and time. Questions about the structure of the universe, about the origin of matter, about the beginning and the possible end of the world, today are not questions of religion but questions to which scientists, with the help of particle accelerators and telescopes, are hoping to find an answer. Modern biology, biochemistry, and biophysics are concerned with the question of the origin of life, and thus ultimately with the problem of our own origin.

Sentient human beings have been asking questions about our origins, about the purpose and meaning of life, since time immemorial. They have sought to find answers in legends, in myths of creation. Every culture created its own myths. Thus the story of the creation of the world, as told by the inhabitants of the Gilbert Islands, says:

Na Arean sat alone in space as a cloud that floats in nothingness. He slept not, for there was no sleep; he hungered not, for as yet there was no hunger. So he remained for a great while, until a thought came to his mind. He said to himself, "I will make a thing."*

And Huai-nan Tzu in the story of creation (China, about 100 B.C.) gives the following account:

Before heaven and earth had taken form all was vague and amorphous. . . . That which was clear and light, drifted up to become heaven, while that which was heavy and turbid solidified to become earth. . . . When heaven and earth were joined in emptiness and all was unwrought simplicity, then without having been created things came into being. This was the Great Oneness. All things issued from this Oneness, but all became different. . . .†

*Carl Sagan, *Cosmos* (New York: Random House, 1980), p. 25.
†Ibid., p. 258.

All stories of creation, including those of the Bible, deal only with the earth, the heavens, and mankind. Since all these myths of creation were thought up by people with very limited views of the processes of nature this is not surprising. Seen from today's perspective these myths explain absolutely nothing. The picture of the world they present mirrors the ideas of the creators of these legends, but it has nothing to do with scientific reality.

The picture of the origin of the world that modern science has given us differs substantially from the myths of creation. Our universe is not only much larger than these myths assumed—it is also far more dynamic; the universe is full of movement, activity. The seeming constancy of the stars in the sky is often linked to stability, permanence. Philosophers such as Aristotle saw this as proof of the eternity of the cosmos, which stood in contrast to the everyday existence of human beings, to the hectic activity and constant changes in our lives.

Science found out that this static, eternal cosmos does not exist, never has existed, and never will exist. Like life on our planet, the universe too is full of activity, subject to constant changes. The stars too are not there for eternity but will disappear some day. New stars are constantly being formed in the universe out of gas and dust clouds, and some end their life in a gigantic explosion.

Some among us may be disappointed to learn that the night sky, that symbol of eternity and stability, was not created for all eternity, that it is only a part of the universe's long evolutionary process of development. I find this new, important insight very satisfying, for it demonstrates that our own existence does not stand in contradiction to the world of stars and galaxies. Just like the galaxies, stars, and planets, we ourselves are part of the process of cosmic evolution. Once we accept this idea, we will find it easier to accept the transitoriness of ourselves as individuals and the transitoriness of our civilization as a whole.

Human history has shown that there are no absolute points of reference. Again and again, millions have fallen victim to illusions that have proved to be unrealistic—whether ideas about the role of gods in natural processes and political ideologies, or ideas about the central role of the planet earth in the cosmos. More than three thousand years ago, at the dawn of civilization, human beings related the phenomena they observed, such as the course of the sun in the sky, clouds moving across the sky, or thunderstorms, to themselves. No natural phenomenon stood on its own; everything was closely linked to human existence and was given meaning only in this interrelation. The view of the world in those early days was highly egocentric. Human beings behaved like little children who believe that the world is created solely for them.

In the course of the past hundred years scientific progress and technological development have produced gradual and mostly unnoticed changes in our thinking. In the past most people had a very simple concept of the world: in its center stood the earth and mankind. The fate of the world lay in the hands of a God who resided nearby and was intimately concerned with the problems of every individual.

We have had to relinquish this simplistic picture of the world. We no longer stand in its center. Scientific research has taught us that we are living at the edge of our galaxy, one among more than a hundred billion galaxies in the universe. All matter, ourselves included, as modern physics has shown us, consists of three elementary building blocks: two quarks and the electron. But even these objects had not been created for eternity. Either the quarks, and with them all atomic nuclei, will disappear in the course of time, or the cosmos will end its life in a gigantic implosion, a reverse of the Big Bang.

Life on earth, modern biology tells us, developed spontaneously, through the constant interplay of chance and necessity. We ourselves are the result of an immense chain of chance developments and events in the dawn of history. We are the result of that history, but we, all of us, are simultaneously the makers of history, enmeshed as we are in the constant stream of events.

The world is an entity, and we ourselves are a minute part of this entity. I see the emerging new view of the world beginning to take shape on the horizon somewhat like this: after centuries of scientific research the fog is beginning to lift. Contours are becoming visible; the entire cosmos lies spread out before us. Many details are still unclear, but today, as we are nearing the end of our millennium, we have reason to believe that we understand the essential features of our universe. As Victor Weisskopf says, "Nature, in the form of man, begins to recognize itself."* Insight and wisdom can assure us that we are on the right path. Without such confidence we cannot long survive on earth.

*Victor Weisskopf, *Knowledge and Wonder* (Cambridge, Mass.: M.I.T. Press, 1979), p. 276.

CHAPTER 2

Galactic Map

> The size and age of the Cosmos are beyond ordinary human understanding. Lost somewhere between immensity and eternity is our tiny planetary home.
> —CARL SAGAN
> *Cosmos*

O N C E while vacationing in Hungary I went canoeing on Lake Balaton. One evening I paddled down the Sio River, which connects the lake with the Danube. The Sio flows through the Hungarian plain for over sixty miles and merges with the Danube near the city of Baja. I had intended to camp along the river shortly after leaving Lake Balaton and to continue toward the Danube the next morning, but I had to scotch this plan. The river banks were too steep, and there were no campsites, so I decided to keep on going. The current was favorable, and I could count on getting close to the mouth of the river by the next morning.

It was an unusually clear, moonless night. Above me was the great span of the Milky Way. The faint glow of the stars guided me through the summer night, lighting my way through the Hungarian landscape. My canoe glided quietly in the calm stream, with only an occasional correction by me to keep my little craft on course, and I was able to concentrate on the stars in the sky.

The Milky Way, the ribbon of stars stretching over the sky from north to south, is our galaxy, a system containing about a hundred billion stars, one of which is our sun —merely one of many stars at the rim of this disk-shaped galaxy. When we look at the ribbon of the Milky Way we are, in a manner of speaking, looking into the disk, that is, along the plane on which the galaxy is spread out.

During my night journey through Hungary I could follow the slow progress of the ribbon across the sky—a consequence of the earth's rotation, like the sun's journey from east to west in daytime. I looked eastward, out of the galactic plane. Here the naked eye can make out some stars in the vicinity of the sun, stars that are still part of our galaxy. But behind that there is nothing but intergalactic space.

In the east, in the Andromeda constellation somewhat above the star Beta Andromeda, I saw a small, diffuse speck of light, the Andromeda nebula. This phenomenon had already aroused the curiosity of the ancients. It is easily seen in southern climes, particularly in the desert regions. In northern latitudes, as, for example, in Central Europe, the Andromeda nebula is visible to the naked eye only rarely, a few times a year at most, and then only in the absence of artificial sources of light such as street lamps.

We now know that the Andromeda nebula is the only visible object in the northern sky that is not part of our galaxy, our Milky Way system. What we are seeing is a "world island" akin to our own galaxy (figure 2.1).

2.1 A view of the Andromeda galaxy. It takes light about 2.5 million years to travel from this, our neighboring galaxy, to earth. Like our Milky Way, the Andromeda galaxy encompasses hundreds of thousands of solar systems, possibly including some inhabited planets. (Reproduced by permission of Hale Observatories, Pasadena, California)

I followed the course of the stars throughout the night, particularly the Andromeda nebula. Like almost everyone else looking at a star-studded sky on a clear night, I had the feeling that this firmament was trying to tell us something. Even though there is nothing more indifferent to our human concerns than the cold light of the stars, still the stars seem to give us something that many of us here on earth are searching for and never find—a firm anchor, something lasting to lean on, a point of reference.

Imagine yourself in a spaceship that is about to leave the solar system. Somewhere in space you see the glittering gas sphere of the sun surrounded by billions of stars, all of them part of our Milky Way system. You realize that you are in a remarkable spot in the universe, namely inside this system, our galaxy.

An observer somewhere in the depths of the universe will as a rule not be fortunate enough to be near such a system. The cosmos is composed largely of empty space; galaxies are rare, precious structures, and observers "residing" in the vicinity of a galaxy can quite rightly consider themselves privileged. We are able to experience the starry sky only because our earth is situated within the Milky Way. Things would be altogether different if the earth together with the solar system were situated somewhere in the universe remote from any galaxy. We would not see a single star in the sky. Instead, we would be enveloped by the eternal night of intergalactic space, and only with the help of complex, powerful telescopes would we be able to see that somewhere far away there exist some faintly shining structures—remote galaxies.

The climate in the depth of intergalactic space is cold and unfriendly, so let's return to our native soil, the Milky Way.

Our galaxy is like the proverbial forest that can't be seen for the trees. Every star visible to the naked eye belongs to our galaxy. Nonetheless it is not easy to gain a precise picture of our galaxy because we are inside it. It would be useful to have a photograph of our galaxy taken from a point outside it. Unfortunately we are not able to send a spaceship equipped with a camera to such a point. If we want to acquaint ourselves with the structure of our galaxy we must try to uncode the structure of the solar system step by step.

We now know that our galaxy looks somewhat like the neighboring Andromeda nebula (see figure 2.1, p. 00).

Our galaxy is disk-shaped, and the sun is located at the rim of this disk. No one has ever counted all the stars in the Milky Way. All we have are rough estimates. According to these figures, our sun has a hundred billion peers, many of them a hundred times more massive, and others smaller.

The stars are not distributed equally in the galactic disk but are concentrated in the so-called arms, spirallike structures extending out from the center. Our galaxy resembles a gigantic wagon wheel with spiral-form spokes. Therefore our galaxy as well as the Andromeda galaxy is called a spiral galaxy (see figure 2.2). That makes sense, for the universe contains galaxies of various shapes.

The stars are not distributed homogeneously throughout the sky; the reason for that is obvious. When we stand at the rim of the galaxy we are able either to look out into intergalactic space or look into the galaxy. In the first case we see only those stars that are accidentally in the vicinity of the sun. In the latter case we see the disk composed of almost a hundred billion stars—the ribbon of the Milky Way. If you happen to be looking at the Milky Way on

2.2 A striking spiral galaxy (Catalogue No. 4 NGC 628). (Reproduced by permission of Hale Observatories, Pasadena, California)

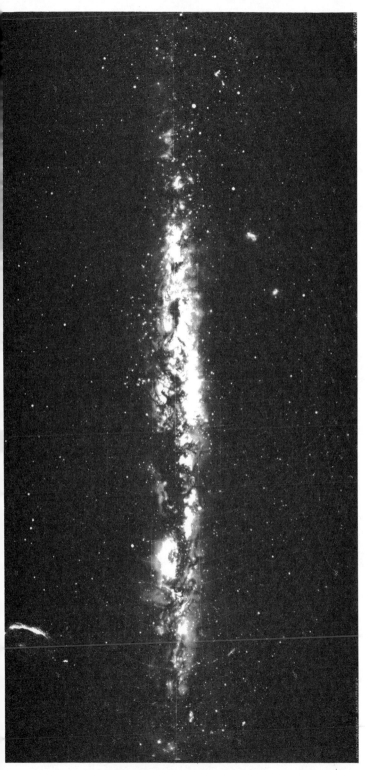

2.3 A panoramic reconstruction of our Milky Way by Knut Lundmark based on actual photographs. On the left, the segment of the galaxy in the constellations Auriga, Perseus, Cassiopeia, and Cygnus. The galaxy's center is located in the vicinity of Sagittarius. On the right, the constellations Centaurus, Crux, Carina, Puppis, and Canis Major, visible only in the Southern Hemisphere. Bottom right, the two Magellanic clouds. At the extreme bottom left, below the Milky Way, the Andromeda galaxy. (Reproduced by permission of Lund Observatory, Sweden)

a clear, moonless night, look toward the constellation Sagittarius. There you will find the center of our galaxy, a region of a vast number of stars, and a vast number of very old ones (see figure 2.3).

Compared to the distances we are used to on earth, the vast distances between the stars in the galaxy defy the imagination. There is little point in expressing astronomic distances in meters or kilometers. The unit of measurement we use is the light-year. A ray of light travels about 300,000 kilometers per second, a distance roughly equivalent to the distance between the earth and the moon. A light signal or radio signal takes about 1/100 (0.01) second to travel from London to New York. The light of the sun takes 8 minutes to reach the earth. Thus the distance between the sun and the earth is 8 light-minutes.

In one year, light travels about 10,000,000,000,000 (10^{13}) kilometers—an enormous distance, and an ideal unit for measuring the distance between the sun and our neighboring stars.* Thus it takes light about 10 years to travel from Sirius, one of the brightest stars in the sky, to earth. Scientists therefore agreed to express astronomic distances in the unit defined by light, in light-years. A light-year is not a unit of time but of distance, namely 10^{13} kilometers. Everyone knows what is meant when the distance between two cities is given in driving time. An hour's driving time is not a unit of time but an indication of the distance a car will cover in an hour (about 60 miles). Astronomers use similar units of measurement, except that instead of a car they use a far more rapid vehicle—light.

There is still another reason why astronomers find the

---

*For a brief discussion of the mathematical meaning of expressions such as 10^{13}, see the appendix.

light-year a particularly appropriate unit of measurement. As I am writing these lines I am sitting in my office in the CERN laboratory near Geneva. On a clear day I can see the peak of Mont Blanc, about 70 kilometers in the distance. When I look at Mont Blanc I don't see the mountain as it looks at that moment but the way it looked 0.0002 seconds before. It takes light that long to travel from Mont Blanc to CERN. This interval is of course minute, and therefore no one would dream of saying, "This is how Mont Blanc looked 0.0002 seconds ago." Instead, we say, "This is Mont Blanc."

In the case of astronomic objects the time factor—the time it takes light to reach us—cannot be ignored. When we are looking at a star, we are not seeing that star as it looks today but as it looked when it sent out the light we are receiving today. Thus the light from Sirius we are seeing now, in 1984, was emitted some years ago, during the presidency of Richard Nixon. And it takes the light of the stars in the center of the Milky Way about 30,000 years to get down to us here on earth. When that light began its long trek, people in Western Europe were still living in caves.

When we look up to the Andromeda constellation, toward the faint glow of the Andromeda nebula, we are seeing light that left that galaxy almost 2 million years ago, at a time when the earth was still uninhabited. There is no way for us to find out how the Andromeda galaxy looks today. The light emanating from its stars now will not reach the earth for another 2 million years.

About 2 million years ago a star exploded in the Andromeda galaxy, near its center. It was what is now called a supernova explosion, a phenomenon in which vast quantities of electromagnetic radiation, much of it in the form of visible light, are sent out. The light produced by this explosion took about 60,000 years to traverse the Andromeda galaxy, and almost 2 million years to travel across the intergalactic space between the Andromeda galaxy and our own Milky Way system. It reached the periphery of our galaxy sometime during the reign of Ramses II, and finally, on August 20, 1885, it arrived on earth, where it was first seen by the astronomer Ernst Hartwig.

Every astronomic sighting, whether with the naked eye or with the help of highly sophisticated telescopes, thus affords us not only a view of remote objects but also a look into the past. The more deeply we look into space the more deeply we look into the past. Many of the stars and systems of stars which we now study with our modern telescopes no longer exist. Some of them have cooled off, and others have ended their lives in great explosions. All that remains of them is their light traveling through intergalactic space.

The light or radio signals leaving our planet today will continue to travel through space for millions of years; and perhaps some day intelligent beings on some remote planet will observe them, at a time when both our solar system and civilization will long have ceased to exist. Possibly someday someone on a remote planet will see our television offerings and conclude that the comparative merits of different brands of soap powders figured very importantly in our lives. For all we know, we may yet become objects of ridicule millions of years hence. Perhaps the inhabitants of a planet in one of the numerous solar systems in the Andromeda galaxy can observe a diffuse smudge of light in the night sky, which just happens to be our own galaxy. They may be able to examine the structure of our galaxy with their instruments, just as, with the help of modern astronomic devices, we can study the structure of the Andromeda galaxy.

At the beginning of the sixteenth century the German astronomer Simon Marius turned his telescope on the Andromeda constellation. "This smudge of light looks like a lighted candle seen at night through a piece of horn," he commented. This simple observation gives no inkling of the vastness of the Andromeda nebula, a gigantic galaxy bigger than our own Milky Way system and containing about 400 billion stars.

We owe much of our present-day knowledge of the structure of the Andromeda galaxy to the German astronomer Walter Baade. During World War II Baade was working at the Mount Palomar Observatory in Pasadena, California. He profited from the war in two ways. A German citizen, he, unlike his American colleagues, was not subject to the draft. Consequently he had virtually unlimited access to the giant 100-inch mirror telescope of Mount Wilson. The observatory was more or less at his disposal. Second, as a precaution against possible air raids neighboring Los Angeles was frequently blacked out; this proved to be a bonanza for Baade, since he could get extraordinarily good pictures of the Andromeda galaxy.

In looking at a picture of the Andromeda galaxy (see figure 2.1, p. 25), we notice that it consists of a very bright central region and a less bright peripheral area. (What we can see with the naked eye or field glasses is the bright center. The outer area cannot be seen without the help of powerful telescopes.) Baade was able to prove that the bright, luminous center region was composed of densely packed stars, that it was not a diffuse, luminous mass as had originally been assumed. He furthermore discovered that the stars in the center were very old, 12 billion years on the average.

Making detailed measurements, Baade also was able to determine that the outer regions of the Andromeda galaxy, like our own galaxy, contain huge gas and dust clouds (particularly clouds composed of thin hydrogen gas). Bright, bluish stars emitting vast quantities of energy in the form of light are often seen in the vicinity of these clouds. Since stars cannot sustain such expenditures of energy for long periods of time, these bluish stars must still be relatively young. It is no accident that they are found in the vicinity of huge dust or gas clouds, for in these clouds stars are continuously being born through the condensation of dust and gas matter.

The advent of radio astronomy, which allowed us to examine the distribution of hydrogen in the Andromeda galaxy, figured significantly in the exploration of its internal structure. A unique property of atomic physics plays a major role in this. Among its various attributes hydrogen gas sends out radio waves with a wavelength of 21 centimeters. Radio telescopes have enabled us to determine the distribution of hydrogen in the Andromeda galaxy. In the process we learned that the hydrogen clouds of the Andromeda galaxy extend far more deeply into space than might have been assumed on the basis of the distribution of the stars. Hydrogen can still be found as far as 100,000 light-years from the center of the galaxy.

As we mentioned earlier, observing with the naked eye or field glasses unfortunately limits our view to the bright central region of the Andromeda galaxy. Just imagine what would happen if the galaxy's faint peripheral region were also visible. Then the Andromeda galaxy would no longer appear to be a small object in the sky; instead, its diameter would be three times that of the moon.

Galactic Map

We now know that our own galaxy is very much like the Andromeda galaxy. To hypothetical observers in outer space our own galaxy and the Andromeda galaxy must look like twins.

The Formation of Stars

Gazing at the stars on a clear night we notice that no two look alike. Not only do they differ in brightness but in color as well. Some emit a reddish light, while others appear to have a bluish tinge, giving evidence that stars are not all identical.

The diversity of the world of stars is indeed impressive; yet the design according to which they were constructed is identical. Stars are composed of the same type of matter as the elements we find on earth, of the quarks and electrons that we will discuss later on. Three quarks combine to form a particle called a proton, the nucleus of the hydrogen atom.

In space, under the influence of gravitational attraction, huge masses of a thin hydrogen gas can contract. In doing so, the gas heats up, and the density and temperature of the gas can ultimately become so great that atomic nuclei fuse. When that occurs great amounts of energy are released, largely in the form of electromagnetic radiation, including visible light. These nuclear reactions take place primarily in the star's interior. Thus the radiation released through this nuclear reaction must make its way through the nuclear matter, a process that can take millions of years. The light we are receiving from the sun today has such a venerable past. It is essentially the product of nuclear reactions that took place inside the sun millions of years ago.

The gravity exerted by a star on its matter tends increasingly to concentrate stellar matter. On the other hand, the radiation emanating from the interior of the star wants to sunder that stellar matter. Equilibrium is established between the two forces, and the conditions of this equilibrium determine a star's size, the color of its radiation, its life-span, and its future fate.*

The stars, including our sun, get their energy through nuclear fusion, that is, through the fusion of atomic nuclei. It is this principle that has found its technical "application" in the hydrogen bomb. The difference between the explosion of a hydrogen bomb and the burning of stellar matter is that gravitational force keeps stellar matter together, and thus stabilizes the combustion process, while in the case of a hydrogen bomb explosion no such stabilization takes place.

We still are not able to control the fusion of atomic nuclei. If this were a simple matter it would be possible to produce almost boundless amounts of energy. But for the time being we profit only indirectly from nuclear fusion, since the "solar energy" which we on earth are receiving either directly (in solar power plants) or indirectly (through the combustion of oil, coal, or gas) is basically energy that originated in the sun through the fusion of atomic nuclei many millions of years ago.

Our sun is a rather ordinary star; our galaxy contains several billions of stars like it. According to the estimates of astrophysicists, our sun is about 4½ billion (4.5×10^9) years old. This figure should not, however, be taken too literally. Astrophysical estimates tend toward uncertainty, since regrettably it is not possible to conduct ex-

*A clear outline of this phenomenon can be found in Rudolf Kippenhahn, 100 Billion Suns (New York: Basic Books, 1983).

periments with the sun. But what's half a billion years! Let us merely say that the sun, and with it our planetary system, including the planet earth, came into being about 4 or 5 billion years ago.

The future fate of the sun can also be predicted. For the next 5 billion years not much will happen. The sun's brightness and size will remain almost unchanged. But thereafter it will begin to increase both in brightness and size. In about 8 billion years the sun will be about a hundred times as big as it is today and will shine about two thousand times as powerfully. It will become a red giant. The sun's inner planets, including the earth, will not survive the birth of the red giant; they will probably be "swallowed up" by the sun.

The end of the sun can also be predicted with a fair degree of accuracy. Ultimately gravity will again condense the matter of the red giant sun. The sun will degenerate into a white dwarf, a highly compressed small star about the size of our earth, in which each cubic centimeter of matter will contain about 1000 kilograms of matter. The surface of such a star is still very hot, hence its name —white dwarf. Finally, this "dwarf sun" will cool off, and the white dwarf will turn into a "black dwarf," a piece of cold matter wandering around the universe. The transformation of the sun into a black dwarf—in other words, the death of the sun—is bound to be a long-drawn-out process, taking an estimated 10^{14} years. There is some doubt whether the universe as a whole will survive that long. That, however, is a problem we will not deal with now.

The sun is merely one among more than 100 billion stars in our galaxy, many of them very much unlike the sun. Some are more than twice its age, while others are extremely young, younger than mankind. The mass of

2.4 The Horse Head nebula in Orion offering a fine view of the grada-
tion from dark to bright dust. The latter is illuminated by "young"
stars. (Reproduced by permission of Hale Observatories, Pasadena,
California)

some stars is many times as great as the sun's, while others have been poorly served by the universe and are smaller than the planet earth.

The space between the stars of our galaxy is filled with dust and gas (see figure 2.4). The density of matter in these interstices is extremely low. Any vacuum produced in a laboratory here on earth contains more matter per cubic centimeter than the average cloud of matter in the galaxy. Yet a great portion of the matter in the universe is nonetheless concentrated in the thin gas or dust clouds. The reason for this is simple: these clouds are enormous. Let us assume we are looking at an area in space in the shape of a die whose edges measure 10 light-years—a rather modest expanse compared with that of an interstellar cloud. Let us further assume that every cubic centimeter of this expanse contains a hydrogen atom. The amount of matter contained in this space can be easily figured out. It is a mass of 2×10^{33} grams, almost the same as the total mass of the sun.

Like the Andromeda nebula, our galaxy also houses bluish stars in the neighborhood of large gas and dust clouds. These stars formed out of interstellar particles as gravity concentrated the matter. Chance can put a large number of atoms or molecules of interstellar matter into a relatively small volume. As a result of gravitation, of the mass attraction between the individual atoms, these atoms will move still closer together. Consequently, still more atoms are captured. This in turn intensifies the attraction of mass, and still more atoms are attracted. This process continues until enough atoms are sufficiently close together to form a star. Thus stars continue to be formed out of interstellar matter. The interstellar clouds are simply gigantic star factories.

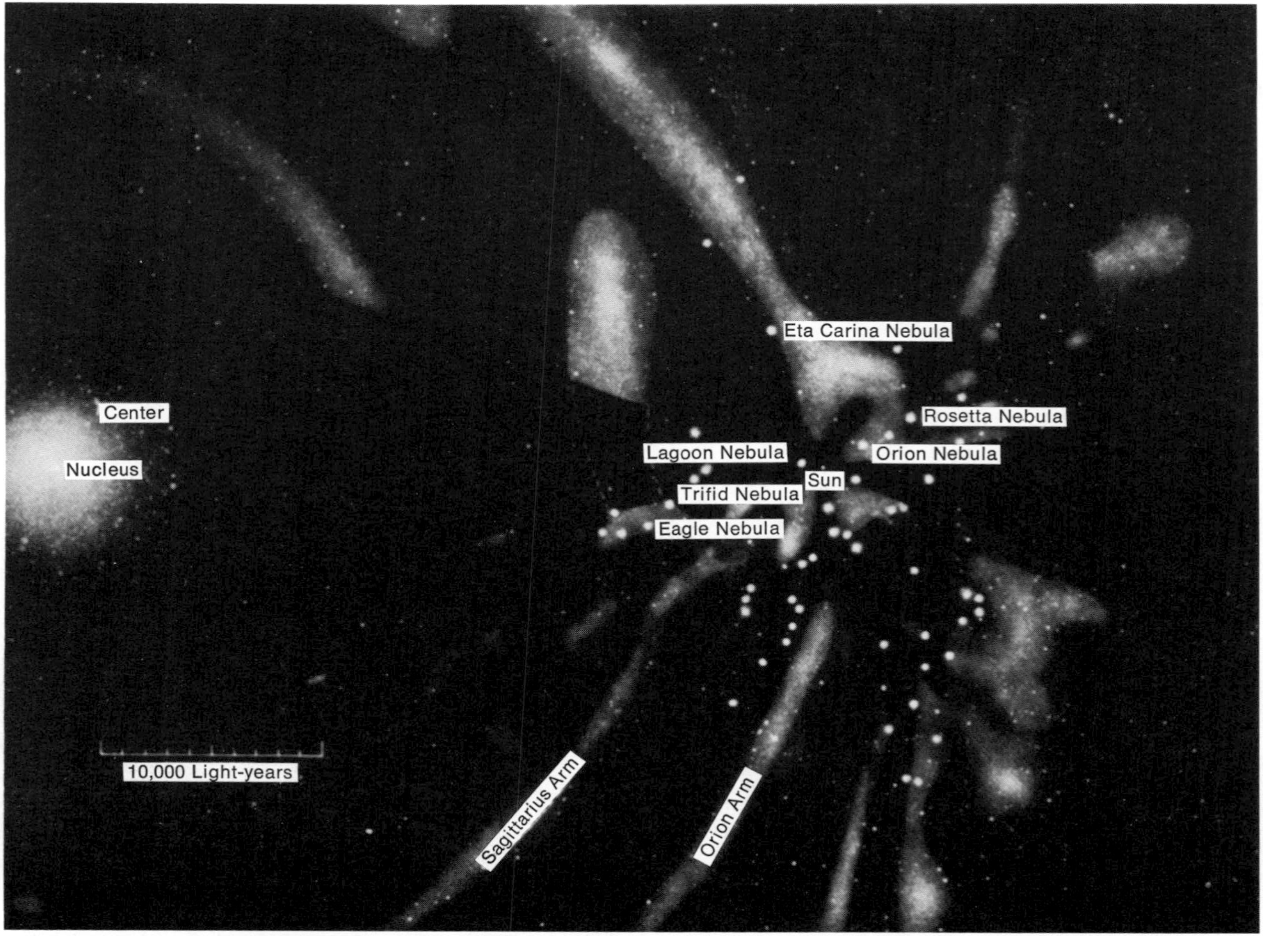

Rosetta Nebula

Orion Nebula

Eta Carina Nebula

Sun

Lagoon Nebula

Trifid Nebula

Eagle Nebula

Orion Arm

Sagittarius Arm

10,000 Light-years

Center

Nucleus

Galactic Map

Black Holes

About 30,000 light-years separate the sun, set amid large gas and dust clouds, from the center of our Milky Way. These gas and dust clouds are part of the galaxy's spiral arms, as for example the arm of Orion or Sagittarius. The sun is located near the inner edge of the Orion arm (see figure 2.5).

Unfortunately the dust and gas clouds near the sun bar our view of some doubtlessly interesting regions. Specifically, we are forever prevented from looking into the center of the Milky Way. We can gain access to these interesting regions only through other means, for example, radio astronomy. Our knowledge about the nucleus of the Milky Way is still not precise enough to tell us exactly what the center of our galaxy looks like and what processes are taking place in it.

Some astrophysicists believe that one or more black holes lie hidden in the center of our galaxy. A black hole is not the figment of the imagination of a practitioner of the occult but a logical conclusion based on the general theory of relativity developed by Einstein soon after the

---

2.5 Plan of a galactic map. The sun lies near the inner edge of one of the spiral arms of our galaxy, named the Orion arm. The conical sections leading away from the sun are not indicative of any particular solar position but rather of the fact that the presence of dark gas clouds prevents us from looking into these regions. The sun is situated near huge gas and dust clouds (nebulae), among them the Orion and the Rosetta nebulae. The dots indicate an accumulation of young stars and/or star clusters. The center of the Milky Way, about which only little is known, can be seen on the left. (Illustration by Sarah Landry from Timothy Ferris, *Galaxies* [Sierra Club, 1980]. By permission of Sarah Landry and Birkhaüser Verlag, Basel)

beginning of World War I. The German astrophysicist Karl Schwarzschild, who died in 1916, a war casualty, discovered shortly before his death that Einstein's equations of gravitation had a simple solution, albeit one with startling attributes.

For example, if we here on earth throw a stone up into the air it will come back down after a given time. The greater the initial velocity of the throw, the longer it takes the stone to return to earth. The reason for this is obvious —the earth exerts attraction on the stone. As a result of this force, the stone slows up until it finally reverses its direction and falls back to earth.

Now let us substitute the beam of a flashlight for the stone. When we direct the light beam directly upward, the following happens: the light waves created by the flashlight travel upward at the speed of light, leave the area of the field of gravity of the earth after a few seconds and that of the solar system after about an hour, and continue to travel through the universe for all eternity. One of the interesting predictions of Einstein's theory of gravitation holds that the light's journey is influenced by the field of gravity of the earth. In its journey the light, because of the gravity exerted by the earth, continuously loses energy (like any moving body, a ray of light also possesses energy, as anyone who has ever taken a sunbath knows). True, this effect of the earth's gravity on light is extremely small and cannot be readily observed. Things become somewhat simpler when we deal with the sun's force of gravity. Soon after the end of World War I an experiment showed that the force of gravity of the sun is able to "deflect light rays from their tracks," and do so precisely in line with Einstein's prediction. This experiment carried out by English astronomers was merely the beginning of a whole series of investigations that lent further credence to Einstein's theory.

The scope of light's deflection by the sun depends on the force of gravity near its surface (or, in the language of physicists, on the power of the gravitational field of the solar surface). The greater this force, the greater the deflection of a light ray (see figure 2.6). Suppose we were

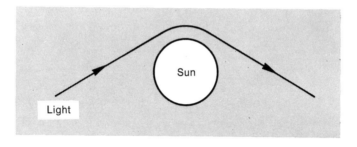

2.6 A light ray, such as the light of a star, being deflected by the gravitational force of the sun. This experiment substantiates a prediction of Einstein's theory of gravitation. (In the diagram the effect is greatly exaggerated.)

able to squeeze solar matter together at will. When the sun's matter is squeezed together its radius becomes smaller, while at the same time the gravitational force at its surface increases. Now let us assume that we are squeezing the solar matter so hard that the sun's radius is reduced to a mere 3 kilometers (this is called the Schwarzschild radius of the sun). In that event, so Einstein's theory maintains, something very peculiar happens. The gravity on the sun's surface increases so much that the light can no longer leave the sun. Gravitation, as it were, calls the light back, much as it affects a stone thrown up into the air on earth. The sun has turned into a black hole from which neither light nor radio waves can escape. A black hole makes an ideal prison.

Do black holes exist in nature? This question is closely linked to the question of whether it is possible for matter to be squeezed together so greatly that a black hole forms. The answer of physicists is an unequivocal yes. Every star possessing sufficient mass (more mass than the sun) will in the course of time turn into a black hole. Our galaxy probably contains hundreds of thousands or perhaps even millions of black holes.

Although black holes are not directly visible, they can nonetheless be found, for example through the gravitation they exert. If our sun were a black hole (which, fortunately, it isn't), the planets of the solar system, including the earth, would revolve around it exactly as they now do.

A black hole is not only an ideal prison but also an ideal waste disposal dump, including an ideal atomic waste dump, since it swallows up all matter in its vicinity. No power on earth could save a space vessel if it accidentally were to come close to a black hole. Its fate would be sealed. Yet even so, matter swallowed up by a black hole does not disappear without warning. Before it is swallowed it heats up and sends out strong X-ray radiation, the death rattle, so to speak, of the captured matter. This radiation and the gravitational force exerted by a black hole help us detect its presence. Since 1970, astrophysicists have discovered a number of candidates for black holes in our galaxy, such as the X-ray source called Cygnus X-1.

Some of the observed phenomena in our galaxy could be understood if one or more very massive black holes were tucked away in the center of the Milky Way. This would have a number of interesting consequences. Since the center of the galaxy contains a great deal of matter—stars and gas and dust clouds—the black hole would be well supplied with food. Like a monster it would gobble up that matter and send out vast quantities of energy in the form of X rays and matter. By the end of this decade we will probably be able to look into the center of the Milky Way with the help of instruments carried by earth satellites. In that case astrophysicists undoubtedly are in for some surprises.

Over the centuries astronomers and astrophysicists have shown us where we are located in our galaxy—at the rim of the Orion arm, about 30,000 light-years from the center of the galaxy, which, like a gigantic wheel, slowly rotates about its axis. It takes the sun hundreds of millions of years to circumnavigate the center of the Milky Way. Since the beginning of human civilization it has not completed even a one-hundredth of this trek.

We are at home here in our galaxy. From here we observe what is taking place around us. We see that our galaxy is only one of many billions. In recent times we have found out where our galaxy is located and how it fits into the community of all galaxies.

But a further question arises regarding our galaxy: where did the matter that formed the stars and planets, dust and gas clouds, and black holes come from? What is matter? Has it always existed or is it continuously being produced in the universe? These questions have occupied scientists for a long time, but only in the 1970s did we learn that answers do exist. In the following chapters we will try to outline these answers.

CHAPTER 3

The Measure of Things

The quantum theory is a wonderful
example of the fact that it is possible to
understand something perfectly yet at
the same time realize that the only way
one can describe it is through pictures
or parables.

—WERNER HEISENBERG
Physics and Beyond

W H E N we look around us carefully we notice that
most things conform to a certain size. For example, the
height of most human beings lies in the 5 foot 3 inch to
6 foot 3 inch range. The leaves of a tree of a particular
genus are of approximately the same shape and size as all
the others of that genus. When we look at salt crystals
under a magnifying glass we see that all of them are
approximately the same size. We notice further that the
size of an object generally does not change in time. Take
a measuring tape or ruler: when you put it away at night
before going to bed and then use it again next morning,
you assume that its dimensions have not changed in the
interim. Why?

Many objects are solid and do not readily change; a steel ball is one object that normally changes neither its shape nor its size. Have you ever wondered why this is so?

Perhaps you have asked yourself what would happen if everything suddenly changed overnight, if all objects increased to twice their normal size? Thus, when you would wake up in the morning you would discover that you were 11 feet 2 inches instead of your usual 5 feet 7 inches. Your bed has also grown by a factor of two, as have your shoes, your rug—everything. Before you say that this does not make sense, that it is pure fantasy, think about it. When you do you will probably arrive at the interesting conclusion that such a spurt of growth in objects could not be detected by any measurements of size alone. How would you know that you are now twice your height? You would measure yourself. But your tape measure has also expanded by a factor of two and thus will indicate your original height, that is, 5 feet 7 inches. Seeking to detect, by measurements of length alone, the expansion of all objects by a factor of two is a futile enterprise—the expansion is unobservable. What does a measure of length mean? When you measure your index finger and see that it is 3 inches long, it means that you are comparing your index finger with a length of 3 inches as indicated on a tape measure. Thus, measuring an object requires the presence of two things: the object itself and a measuring device. There is no such thing as absolute length. Whenever we want to know the length of an object we must compare it to another object, a measuring gauge. When we blow up a balloon, for example, we change its dimensions compared with an established standard, which of course does not change when we blow up the balloon.

Let me now pose another question, one you may consider to be as strange as the one about the expansion of objects. The toothbrush you use at night is about as long as your open hand. Have you ever wondered the next morning why your toothbrush is still the same length and not, for example, twice as long—or perhaps as small as a coin? Probably not. Things obviously do not change overnight. But why not? What is the reason for the stability of objects?

How Big Are Atoms?

People have asked themselves questions like these for thousands of years without finding fully satisfactory answers. The first attempts to formulate answers were made by the ancient Greeks.

The 100-drachma Greek banknote has on it the likeness of Democritus, one of the most original thinkers of ancient Greece. This man, who anticipated the ideas of modern physics, was a native of the city of Abdera in northern Greece. He was an unusual philosopher, at least by the standards of modern philosophy. Democritus considered the enjoyment of life the highest good, and enjoyment for him was identical with understanding. He coined one of the most remarkable phrases of any Greek thinker: "Nothing exists except atoms and emptiness." He postulated that atoms are the smallest indivisible constituents of matter. We can cut an apple in half, said Democritus, only because the apple must consist of minute indivisible constituents separated by nothing but empty space, so that a knife can cut between the atoms, through this empty space. Democritus further concluded that objects are stable and do not change over time because all atoms are of equal size. The atoms themselves are unalterable and indivisible. The differences we find between substances—such as iron being harder than wood—occur because the atoms in different objects are arranged differently.

Take the letters of the alphabet. By arranging them in certain patterns we can express a variety of different ideas, write a potboiler or a newspaper article or a poem. What matters is the way the letters are arranged.

Democritus was at least partially right in his assumptions. At the beginning of the nineteenth century the atomic idea dominated chemistry. The chemical properties of materials, chemists discovered, could be understood only if the materials were composed of the minutest units—atoms. These were assumed to be small spheres with a radius of about 10^{-8} centimeters. To visualize the minuteness of atoms let us imagine that each of the one billion inhabitants of the Peoples Republic of China is given one atom. If all these atoms were put next to each other to form a chain, that chain would only be slightly longer than your middle finger, about 4 inches.

Toward the end of the nineteenth century some physicists began to question whether the atoms of substances studied by chemists were in fact similar to the atoms postulated by Democritus. Their skepticism was linked to the strange results of experiments carried out in Paris, which showed that certain atoms apparently were not stable but were sending out particles, most likely constituents of atoms. The phenomenon of radioactivity had been discovered.

About eighty years ago Ernest Rutherford in England

made a discovery that may well be the starting point of modern atomic physics. It already was known that some radioactive atoms emit particles called alpha particles. At this point it does not matter to us what these alpha particles signify or what they are composed of. What is important is the fact that they are being hurled out of many unstable atoms at relatively high speed, almost like small projectiles; that is precisely how Rutherford used them. Through them he sought to explore the interior of atoms. He let the alpha particles fly through a thin sheet of metal (he used gold) in the hope that he might occasionally see a deflection of these particles from their original direction and thereby draw conclusions about the structure of atoms. If the matter in the interior of atoms is distributed more or less uniformly one would expect that an alpha particle traveling through an atom would alter its flight line and speed only a little, but not dramatically. Compare this with a bullet going through a bag of water. The resistance of the water would brake the bullet but would not change its direction abruptly. The latter could happen only if the bullet came up against a solid object, perhaps a piece of iron in the water (see figure 3.1).

To his surprise, Rutherford found that some of the alpha particles did not penetrate the metal sheet but were deflected backward. He said that this was the most incredible event he had ever witnessed, as incredible as if a 15-inch shell fired at a piece of tissue paper came back and hit the person firing it.

After a series of experiments Rutherford arrived at a remarkable conclusion: the recoil of some of the alpha particles could be explained only if one assumed that practically all matter in an atom is concentrated in a volume much smaller than the atom itself—in an atomic nucleus.

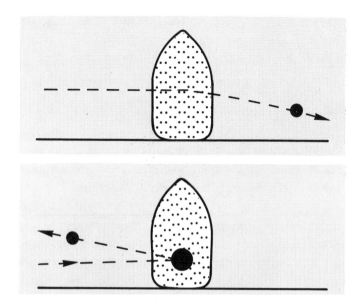

3.1 A marksman aiming at a bag of water. The bullet passing through the bag is slightly deflected and braked by the resistance of the water; its flight track, however, does not change abruptly. But if a chunk of iron is present the direction might change; should the marksman hit the iron he runs the danger of being hit by the deflected bullet. If we substitute an alpha particle for the bullet, an atom for the bag of water, and an atomic nucleus for the chunk of iron, the diagram illustrates Rutherford's experiment.

The diameter of the atomic nucleus ranges between 10^{-12} and 10^{-13} centimeters, which means that it is ten thousand times as small as the atom itself. If a billion atomic nuclei were put end to end they would form a chain less than 1/100 (10^{-2}) millimeter long.

What, then, determines the volume of an atom when most of its matter is concentrated in the much smaller nucleus? The atomic nucleus is not an isolated entity. It is surrounded by a shroud composed of electrons—particles much smaller than either atoms or atomic nuclei.

These electrons revolve around the atomic nucleus. Their mean distance from the nucleus is about 10^{-8} centimeters, and that distance determines the size of atoms. The atom is something like a small planetary system, with the sun (the atomic nucleus) in the center, surrounded by planets (electrons) that revolve around it.

To get a sense of the comparative sizes of the structural elements of an atom, imagine an apple set in the center of a baseball diamond as the nucleus; the electronic paths would extend somewhere near the bleachers.

Thus, as this example suggests, atoms are practically empty. Only a minute fraction of their volume contains matter. In the structure of matter nature treats empty space wastefully. If all the atomic nuclei and electrons on earth were compressed into the smallest possible space we would get a sphere with a radius of about 100 meters.

In the preceding chapter I mentioned that black holes can form only if matter is highly compressed. The mass of the sun would have to be squeezed together in a sphere with a radius of 3 kilometers. This is not impossible, provided all atomic nuclei are greatly compressed. A black hole thus is an altogether normal object, completely consistent with scientific knowledge. To doubt the existence of black holes is tantamount to doubting the known laws of nature.

What compels electrons to revolve around the atomic nucleus? We know that the earth revolves around the sun because of an attraction exerted on the two bodies: gravitational force. It is, however, also clear that the gravitational force is far too weak to force electrons to revolve around an atomic nucleus. The motion of the electrons in the atom can be understood only if a different attractive force is at work between the electrons and the atomic nucleus. What is the nature of that force?

Electric Charge

If you walk over a rug and then touch a doorknob, sparks sometimes fly between the doorknob and your fingertips. There is a simple explanation for this. The atoms that make up the rug are able to shed electrons easily. When your shoe touches the rug some of the electrons cling to your shoe. These electrons possess a characteristic known in physics as their electric charge. When the electrons collect on your shoe they produce an electric charge that is transferred to your body. Your entire body becomes electrically charged. The charge of the electrons is negative. ("Negative" is a traditional definition, not a pejorative description, of the electronic charge; it was first introduced in the eighteenth century by Benjamin Franklin.) When you walk across a rug your body may acquire a negative charge. Under normal conditions a rug is not electrically charged; it is neutral. However, if it suddenly sheds electrons it lacks the previous negative charge and thereby acquires its opposite: a positive charge. When you then touch the doorknob a neutralization of the electric charges takes place. The electrons on your body leap over to the doorknob, producing a spark that can even be a little painful. The leap of the spark neutralizes the charge. After you leave the room neither your body nor the rug will be electrically charged—everything is in neutral balance.

It has long been known that electrically charged bodies interact. Two bodies carrying the same type of charge, either negative or positive, repel one another. Dissimilar charges, however, attract each other. How these forces originate will be discussed later on.

Electric forces of attraction and repulsion play an important part in the structure of atoms. The atomic nucleus carries a positive charge. Consequently there exists an electric attraction between the nucleus and the electrons, and it is this force that keeps the atom together. Normally an atom is electrically neutral; it possesses no net electric charge. If we take away an electron from a neutral atom the positive charge of the nucleus predominates. The atom is now positively charged.

This explains why we become negatively charged when we walk over a rug. One might think that the charge could be negative one time and positive another time, depending on the prevailing conditions. But one never obtains a positive charge in this situation, and the reason for this is simple. The negatively charged electrons from the rug's atoms can easily be brushed off, but not so the positively charged atomic nuclei. They are firmly anchored in the atomic structure of the atoms of the rug.

Two negatively charged electrons repel each other. This, too, plays an important role in the structure of matter. When you put a pencil down on the table it will rest on the tabletop and not fall through the tabletop to the ground, even though the earth exerts attraction on it. The pencil would like to fall, but the tabletop offers resistance. Why? The reason is relatively simple. When the pencil rests on the table, the atoms of the pencil and those of the tabletop are very close together. Their atomic shells practically touch. This means that the electrons of the pencil and of the tabletop are fairly close, and they repel each other. This repulsion manifests itself as the resistance of matter. Were it not for the repulsion of the atomic shells, the pencil would fall through the tabletop.

This demonstrates that the electric forces of attraction and repulsion are of prime importance in the structure of matter. If the electric forces in the world were turned off (fortunately this cannot be done) everything would immediately turn to dust composed of atomic nuclei and electrons.

Many different types of atoms exist. What distinguishes them from each other is above all the number of electrons in their shell. The hydrogen atom is the simplest atom, possessing only a single electron. Its atomic nucleus is also very simple, consisting of one positively charged particle called the proton. Compared to the electron, the proton is a rather massive particle. Its mass is almost two thousand times that of the mass of the electron. Later on we will discuss the proton in greater detail. The electron, unlike the proton, as we shall see, is an elementary object, which means that it does not consist of smaller parts. (At least so far, nobody has found evidence that electrons are composite objects.)

The shells of more complex atoms possess a number of electrons, sometimes quite a substantial number. One might suppose that the nuclei of these atoms are composed of nothing more than of a correspondingly large number of protons—and that is what physicists originally believed. Rutherford, however, did not hold with this idea. He believed that the atomic nucleus had to contain still another kind of particle, a neutral particle, which he dubbed the "neutron." Protons and neutrons are the constituent parts of the atomic nucleus, and the nucleus, Rutherford conjectured, like the atom itself, is a divisible object, not an atom in Democritus' sense.

He was to be proved right. In early 1932 one of his co-workers at the Cavendish Laboratory at Cambridge, James Chadwick, discovered a strange radiation composed of particles that were not electrically charged and that had a mass almost equal to the mass of the proton. The neutron had been found.

Since neutrons are electrically neutral it is not possible to influence neutron radiation through electrical fields. They consequently penetrate matter with relative ease, since the strong electric fields of the atomic nuclei present no obstacle to them. Only the rare direct collision of neutrons and atomic nuclei can (after a considerable time) stop neutron radiation.

We now know that atomic nuclei consist of protons and neutrons. The number of protons in the nucleus corresponds exactly to the number of electrons in the shell. For example, the gold atom contains 79 electrons, 79 protons, and, almost always, 118 neutrons.

A Smallest Unit of Charge

How big is the electric charge of an electron? Do all electrons carry identical electric charges? In 1910 the American physicist Robert A. Millikan, a native of Iowa, asked himself these questions (see figure 3.2). As a student Millikan had been noted for his keen intelligence but showed no great interest in the natural sciences. Languages, classical Greek in particular, were his true love. By coincidence, his Greek instructor also taught physics, and one day he asked Millikan to take over an elementary physics course. Feeling that physics was not exactly his forte, Millikan was reluctant to do so, but his teacher maintained that anyone who could master Greek could teach physics. And so Millikan found himself instructing the class. He became so fascinated with the subject that he decided to make physics his life's work. His Greek teacher's offhand remark turned out to be of crucial importance not only for

3.2 Nobel Prize laureates Albert Michelson (left) and Albert Einstein (center), with Robert A. Millikan (right), at a conference in Pasadena, California, in January 1931. At this conference Millikan vainly sought to persuade Einstein to come to the United States. Where Millikan failed, Hitler succeeded. Soon thereafter, Einstein left Germany for good. His emigration signaled the collapse of German science in the Nazi era. (Courtesy of the Archives, California Institute of Technology, Pasadena)

Millikan's future, but for the course of science in the United States as well.

For his contribution to his chosen field Millikan received the Nobel Prize for Physics. He was the first scientist to have his picture appear on the cover of *Time* magazine. In 1921 he joined the faculty of California's new Institute of Technology, soon to become one of the most important centers of physical research in the U.S. "Just imagine," the German physicist Wilhelm Roentgen is reputed to have said when he heard of Millikan's appointment, "they say that Millikan has $100,000 a year to support his research." (In the twenties, $100,000 was looked on as a princely research grant.)

To determine the electric charge of electrons, Millikan measured the charges on minute droplets of oil. Either the atoms of such droplets contain an equal number of electrons and protons (in which case the droplets are electrically neutral), or the number of electrons is greater or smaller than the number of protons (in which case the droplets are negatively or positively charged, respectively). Millikan found that the charge of the droplets he tested was invariably a whole-number multiple of a specific charge, which he dubbed the elementary electric charge. Its numerical value has since been precisely established, but does not concern us here. The charge of the droplets, for example, turned out to be two or three times that of the elementary charge. Obviously the elementary charge was simply the charge of an individual electron.

The atom of the element hydrogen consists of one electron and one proton. We know that the electric charge of a hydrogen atom is zero. Consequently, the charge of the proton must be exactly equal to the charge of the electron. The two differ only in their respective sign. It is also obvious that even in a complex atom the number of electrons in the shell is equal to the number of protons in the nucleus, for only in that situation can the charges of the electrons and protons cancel each other to make the atom as a whole neutral.

The fact that the electric charges of electrons and protons are identical, except for their signs, puzzled physicists from the beginning, for in all other respects electrons and protons are very different sorts of particles—as their mass alone indicates. Protons are almost two thousand times as heavy as electrons. One might expect that their charges also would differ, even if only slightly.

The identical size of their respective charges probably

signifies that protons and electrons share a characteristic —that something links them together. Physicists long searched for such a relationship without the slightest success. Only recently was light shed on this enigma; the medium of this enlightenment was the quark model of protons, which we shall consider later.

The Size of Atoms

At this point let us look at another aspect of atomic physics: the size of atoms, more specifically the size of hydrogen atoms. As mentioned earlier, hydrogen atoms are the simplest type of atoms, consisting of only a single proton and a single electron. It is relatively easy to make a hydrogen atom. All that is needed are the requisite building blocks, a proton and an electron. Let us assume that we have a proton and electron at hand. Since they attract each other electrically, building a hydrogen atom presents no problem. All we have to do is bring the two particles into close proximity; the rest happens automatically. They fly toward each other and combine, forming a hydrogen atom, and in the process they emit energy in the form of electromagnetic radiation (such as, for example, light).

Now what about the size of the atom? In principle we can imagine that the electron and proton are relatively far apart, about 1 centimeter, or very close, let us say far less than 10^{-10} centimeters. As mentioned earlier, the normal size of an atom is about 10^{-8} centimeters. What determines its size? Are all hydrogen atoms equally big? Experiments have shown that there exists a critical size, namely 10^{-8} centimeters. No hydrogen atom can be smaller than this, and although it can be larger, it cannot remain so for long.

If we make a hydrogen atom in a laboratory out of a proton and an electron, after a short time that atom will assume its typical size—which is to say, the orbit of its electron will have a radius of about 10^{-8} centimeters. Let us assume our colleague in the adjacent laboratory is also "building" a hydrogen atom. When we compare the two finished products we find that the atoms are exactly the same size.

To pursue this example still further, let us assume that astronomers will someday succeed in establishing radio contact with an extraterrestrial being on a remote planet, a possibility that cannot be dismissed out of hand. Among the many questions this alien creature is sure to ask probably will be one concerning the size of planetary beings. We would promptly radio back that the average height of an adult earthling is about 5 feet 9 inches. The next question we can expect to be asked will concern the length of an inch, for how are our neighbors in the universe supposed to know that? We then remember what we know about the structure of the atom and answer, "On the average our body length is 3.5×10^{10} times the radius of a hydrogen atom (equivalent to about 5 feet 9 inches)."

Since we may assume that our neighbors in the universe know as much about the structure of atoms as we, this information will mean something to them, and they will be able to figure out the size of our unit of length. They in turn might tell us that they are somewhat taller or shorter than our average.

The size of atoms was a problem that occupied physicists in the first two decades of our century. A solution was found in the middle of the twenties—not a simple one, but one that changed our basic ideas about the processes taking place inside atoms. A new world was discovered, the world of the quantum.

Why Are Atoms Stable?

In the summer of 1922 two men—one a tall middle-aged scientist and the other a student—were walking together in the woods surrounding the university town of Göttingen. The first was a visiting physicist from Copenhagen, Niels Bohr, and the other a graduate physics student from Munich, Werner Heisenberg. Bohr talked of nothing but atoms and their size. He thought it miraculous that all hydrogen atoms were of equal size; he called this phenomenon the miracle of the stability of matter. Many years later Heisenberg recalled this conversation and paraphrased Bohr's discourse thus:

By "stability" I mean that the same materials with the same properties occur again and again, that the same crystals are formed, that the same chemical combinations come into being, and so forth. That must mean that even after many changes brought on by external effects, an iron atom ultimately remains the same iron atom with the same properties. According to classical mechanics that appears incomprehensible, particularly when an atom resembles a planetary system. In nature there thus exists a tendency to fashion definite forms . . . and to re-create these forms even when they have been disrupted or destroyed. In this connection one might even think of biology, for the stability of living organisms, the creation of the most complex forms that are only temporarily viable entities, is a similar type of phenomenon. But in biology we are dealing with highly complex, temporally changeable structures which are not at issue here. I am here concerned only with simple forms with which we are familiar from physics and chemistry. The existence of uniform materials, the presence of solid bodies, everything depends on the stability of atoms.*

This conversation with Bohr was to have a profound effect on Heisenberg. Upon his return to Munich, Heisenberg continued to work on the structure of the atom under Arnold Sommerfeld. Again and again, his thoughts went back to the problems touched on by Bohr during their walk. And they continued to occupy him even after he returned to Göttingen in the summer of 1924 and joined Professor Max Born's seminar in theoretical physics. Here in Göttingen he finally solved the problem of atomic stability with a solution that demanded a radical shift in thinking, the discarding of the classical concepts that had dominated physics since the days of Newton—the sort of shift that probably only scientists as young as Heisenberg are able to make. The quantum theory was born (figure 3.3).

The Quantum

The birth of the quantum theory represents a milestone in the history of science. It became apparent that in the microworld of atoms the concepts we had used to describe our macroscopic world—traditional ones like the speed of an object—had to be interpreted differently than in our familiar macroscopic world. The quantum theory not only brought a revolution into the world of physical concepts, it also changed our ideas about the cosmos as a whole. Its

*Werner Heisenberg, *Der Teil und das Ganze* (Munich: Piper Verlag, 1969), p. 285. [Translated in English as *Physics and Beyond* (New York: Harper and Row, n.d.)]

3.3 The thirty-three-year-old Werner Heisenberg (standing) with Niels Bohr in the Bavarian Alps in 1934. (Courtesy Heisenberg Archives, Max Planck Institute, Munich)

implications for other areas of intellectual life such as philosophy are still not thoroughly understood. I find it strange that no wide-ranging discussion about quantum phenomena—comparable to the discussion carried on since the twenties about the significance of Einstein's theory of relativity—has gotten under way.

Quantum theory seems to be considered a kind of secret science of physicists, far too abstract to be made accessible to the broad public. Admittedly it is an abstract affair. One undoubtedly would find it impossible to understand all facets of the atomic structure without a grounding in the mathematics of the theory. Still, I believe that nonphysicists and nonmathematicians can understand the basic ideas of quantum physics. I even believe that it is essential that they do. For decades, quantum physics was limited to physical research and nothing else. That, however, is no longer the case, not only because of the atomic bombs dropped on Hiroshima and Nagasaki and the development of nuclear technology, but largely because of the development of electronics. Every television set contains elements that utilize the phenomena of quantum physics. Quantum physics plays a role in microcomputer technology. Before long the electronic circuits in microcomputers will be so minute that the effects of quantum mechanics will become a significant factor. Many industrial enterprises are dependent on quantum physics. Is it unreasonable to demand that engineers, physicians, and other professionals join physicists, mathematicians, and chemists and learn something about quantum mechanics?

What were the major ideas that helped physicists working with Heisenberg at Göttingen solve Bohr's problem of the stability of matter? The general task that physicists and scientists set for themselves is to determine relationships between observed quantities: if a given quantity, let us say the location of an object, is such and such, then this and that follows. This sort of task applies not only to the natural sciences but to all branches of knowledge—economics, for example, which is engaged in the study of both quantitative and qualitative characteristics. There exists a correlation between interest rates and unemployment. A drop in interest rates stimulates industrial investment, and as a consequence unemployment goes down. In contrast to physics, the relationships between observed quantities in economics are generally multilayered and complex, and as a result the predictions of economists frequently (and perhaps understandably) are not accorded much more credibility than the predictions of soothsayers and astrologers.

What is peculiar about quantum mechanics (as quantum physics is called) is the fact that it is no longer simple to determine what an observed quantity really is. Let us consider the hydrogen atom. As we know, this simplest of all atoms consists of two particles, a proton and an electron. Since the proton is far heavier than the electron, one can in fact imagine the proton as being at rest, with the electron moving in an orbit—let us say a circular orbit—around the nucleus, that is, the proton. Now one probably will ask, How do you know that the electron moves in a circular orbit? Have you seen it?

In trying to answer this question it is easy to get flustered. Of course no one has ever actually observed an electron in motion. All we know is that the hydrogen atom consists of one proton and one electron, and that the electron somehow moves around the proton. And if it does, then why not in a circular orbit? Since we are confronted with this problem let us examine it in greater detail. In order to find out whether the electron does in fact move in a circular orbit, we must measure its position at different times. How do we go about this?

Measuring Position and Speed

Let us leave the atom for the moment and talk about this type of measurement more generally. Assume that we want to determine the position and velocity of a moving automobile. That is simple. We use a device popular with highway patrols: a radar trap. Many of us know all about them. If you should be unlucky enough to be caught in one, you'll probably get a ticket saying that on this particular day at such and such a time you were exceeding the speed limit by so many miles per hour on this specific stretch of highway. The summons records the time, place, and speed of the moving vehicle.

How is this speed measured? A radar station sends out radar waves, a special type of electromagnetic wave. These hit your car, are reflected by it, and are recorded by a receiver. The radar waves are scanned by the receiver, making it possible to record precisely the speed at which your car was traveling. By the same token, the location of the car at the time the radar waves were reflected can also be determined by measuring the travel time of the waves.

Thus we see that signals, in this case radar signals, play a role in fixing both the location and speed of the car. These signals must interact with the object under observation (in this case the radar waves reflecting off the car). It turns out that signals represent a form of energy. Radar signals are a vehicle for transporting energy. Radar waves, like light, are simply electromagnetic waves. When we expose ourselves to the sun, the sun transports energy to our body, above all to the skin. Radar waves act in like fashion. When the signal hits the rear of the car, the car receives bits of energy and impulse, although in minuscule amounts. The car accelerates but so imperceptibly that this acceleration can be ignored. Therefore, by ignoring the effects of the signals used to measure speed and location, we can say that the speed and location of an object like a moving car can be measured without any difficulty.

However, in the case of measurements of a very small object, such as an electron, difficulties do arise. Let us assume we want to measure the location and velocity of the electron in a hydrogen atom at a specific point in time. First, we must decide how to carry out the experiment. Like the traffic police we build a sort of radar trap, that is, we use electromagnetic waves (perhaps light waves) in our measurement. We are able to determine the position of the electron by irradiating the atom with light. The light waves are reflected by the electron thereby permitting us to draw conclusions about the position of the electron at the time of reflection.

How do we go about measuring velocity? When light waves are reflected by the electron, they transfer an impulse to the electron, thereby changing its state of motion in an unpredictable manner. Determining both its precise velocity and its location becomes impossible. The quantities *location* and *velocity,* or location and impulse, are said to be mutually complementary. (The concept *impulse* is here used in its physical sense; the impulse of an object is simply its mass multiplied by its velocity. At identical velocities, the impulse of a truck is thus substantially greater than the impulse of a passenger car.)

Uncertainty

Heisenberg was the first to recognize the significance of the complementarity of two physical quantities. He found that there exists a clear relationship between the levels of certainty or uncertainty of measurable physical quantities; he called it the uncertainty relationship. The uncertainties of physical quantities are determined by a constant discovered by the German physicist Max Planck in the early days of this century. Planck's constant, named for its discoverer and usually designated by a slashed $\hbar$, is a constant of nature; its value is given as

$$\hbar = 1.05 \times 10^{-34} \text{ (kg)m}^2/\text{sec.}$$

It is not a simple number, but one composed of units of kilogram (meter)^2 per second—a strange unit indeed, but an appropriate one, since Planck's constant describes a quantity we do not ordinarily deal with, the *action*.

What do Heisenberg's uncertainty relationships tell us? As far as the complementarity of location and impulse is concerned, they tell us that the uncertainty of the impulse multiplied by the uncertainty of location cannot be smaller than $\hbar$. If we designate the uncertainty of the impulse of an object as Δp and the uncertainty of its location as Δx, we arrive at the relationship

$$\Delta p \times \Delta x \geq \hbar,$$

Expressed in words instead of symbols, this relationship tells us that the product of the uncertainties of impulse and location must always be greater than, or equivalent to, $\hbar$.

Expressed in the above units, Planck's constant is a very

small number, and that is why the effects of quantum theory can be ignored in our daily lives. As an example let us take the quantum-mechanically caused uncertainty of the velocity of a moving car weighing 1000 kilograms. Let us assume that the highway patrol is able to fix the location of the car within 1 centimeter, which gives us

$$\Delta x = 1\text{cm} = 10^{-2}\text{m}.$$

With the help of Heisenberg's principle we thus arrive at $\Delta p = \hbar / \Delta x$. We know that the impulse is the product of mass and velocity. Applying this relationship we can find the uncertainty of velocity: $\Delta v \approx 10^{-35}$ meters/second. This uncertainty is so inconceivably small that we can safely forget all about it. At any rate, you would not be able to persuade a judge that you are not guilty of speeding because the traffic cop failed to take the quantum-mechanic velocity uncertainty into account when he gave you that speeding ticket.

Matters, however, look entirely different in the case of an electron. The mass of an electron is very small, only 9.1×10^{-31} kilograms. Even though a liter of water contains about 3×10^{25} electrons, they add very little to the total weight of the water, only about 1/3 gram; the atomic nuclei are responsible for most of the weight.

Because of the smallness of the electron mass, the uncertainties in the determination of location and velocity are at times considerable. When we succeed in determining the location of an electron within a hundredth of a millimeter, the uncertainty in determining its velocity amounts to 10 meters/second, or 36 kilometers/hour. If a car had the mass of an electron, a quantum traffic cop would have a problem in enforcing speed limits. Violators could invoke Heisenberg's relationships and call in physicists as expert defense witnesses, thereby opening up the sort of lucrative opportunities for scientists that have been the traditional preserve of lawyers.

What makes the uncertainty principle so important is its role in establishing the size of atoms. A physicist knowing nothing about quantum mechanics would undoubtedly be puzzled by the stability of atoms. Not only would he be astonished to learn that all hydrogen atoms are of identical size, about 10^{-8} centimeters in diameter, but he would also be puzzled by another problem. Let us assume that the electron in a hydrogen atom moves in a circular orbit. The electron carries an electric charge. We also know that an electrically charged object moving in a circular orbit constantly sends out electromagnetic waves, somewhat like a small radio transmitter. Since electromagnetic waves, including light, are simply a special type of energy, the electron in the hydrogen atom would be losing energy constantly. That energy would have to come from the electron's orbital movement, and consequently the electron would move closer and closer to the proton in a sort of spiral maneuver, until finally it would fall into the atomic nucleus. All this, of course, stands in contradiction to what is actually taking place.

We can easily figure out how long a hydrogen atom would live before its nucleus swallowed up the electron. The answer is astonishing: less than one second. We find, however, that hydrogen atoms live far longer than a second. As far as we know, hydrogen atoms are stable—that is, they live for infinite periods of time (except for some exotic processes such as proton decay, which we will discuss later). What causes the electrons in a hydrogen atom to revolve evenly in their orbit?

The answer is found in the uncertainty principle. Let us

examine the hydrogen atom more closely. The uncertainty of the location of the electron is determined by the size of the atomic shell, which has a diameter of about 10^{-8} centimeters. Let us assume that we can find a hydrogen atom with a far smaller atomic shell, say, one of about 10^{-11} centimeters in diameter. In such an atom the electron would be better localized than in the usual hydrogen atom. On the basis of the uncertainty principle, this means that there is far greater uncertainty of impulse in the smaller atom, and consequently also greater uncertainty of velocity. In the smaller atom the electron will move more rapidly than in the larger one. But greater speed means greater energy, for the more rapidly a body moves the greater its energy. It would thus appear that the electron in the smaller atom has more energy than its counterpart in a typical atom. According to an important principle of nature, every system seeks to exist in a state of lowest energy. The smaller atom would thus not be "viable." Consequently, by emitting electromagnetic waves, it would seek to expand until it reached the size of the ordinary atom.

Now let us look at a hydrogen atom much larger than its ordinary counterpart. Let us assume it has a diameter of 10^{-4} centimeters. The electron in this atom would be farther away from the nucleus than in an ordinary atom. To construct such an atom we would practically have to pull the electron away from the nucleus. That is to say, we would have to expend energy to make such an atom, which means that our big atom again has far more energy than the usual hydrogen atom.

Whatever we do, nothing seems to work: the ordinary hydrogen atom possesses the lowest energy. The rest, all our artificial creations, have more energy and therefore are not stable, and after the emission of electromagnetic waves (for example, light), promptly turn into typical hydrogen atoms.

The typical hydrogen atom therefore has a characteristic diameter of 10^{-8} centimeters, simply because it does not pay for the atom to be either larger or smaller. Due to the uncertainty relationship, it "sits" in a state of lowest energy. No power in the world can force the electron in the hydrogen atom to give away energy. Quantum theory does not permit it. With the help of the quantum theory the radius of the hydrogen atom can be figured out exactly; it depends on Planck's constant $\hbar$ and on the electron mass. The result, in perfect accord with the experiment, is a quantity known as Bohr's radius, namely 0.53×10^{-8} centimeters.

Bohr's radius is one of the basic quantities of nature. Quantum theory has shown us that in their normal state —their state of lowest energy—all hydrogen atoms in the universe, whether here on earth, in intergalactic space, or in distant galaxies, are of equal size. Quantum physicists therefore find it useful to measure length in units of Bohr's radius. Let us call this unit a *Bohr*. One meter would thus be equivalent to 1.9×10^{10} Bohr. The average height of an adult would be 3.3×10^{10} Bohrs, and so forth. Of course this unit would not be particularly useful in our daily lives, for we ordinarily deal in far larger units. Nonetheless, the Bohr, nature's unit, makes more sense than our traditional inches and feet and meters, which have no fundamental physical meaning.

The analysis of the properties of hydrogen applies to all atoms. In contrast to the hydrogen atom, more complex atoms contain more than one electron, some of them even a considerable number. But every atom has its own predetermined state of lowest energy. Left alone, every atom will in time arrive at that state, called the ground state. Thus, as in the case of the hydrogen atom, quantum theory fixes the size of atoms. Other properties of atoms are also fixed. For example, an atom containing thirteen electrons is invariably the atom of the metal aluminum, regardless of whether it is found here on earth or on the moon. An atom with ninety-four electrons is invariably that of the element plutonium.

Quantum theory even allows us to figure out the chemical and physical properties of a given substance. Take plutonium, which is the product of nuclear reactions. Plutonium does not exist as a stable element on earth since it disintegrates into other elements in the course of 40,000 years. Physicists, analyzing the various properties of plutonium, concluded that plutonium had to have the properties of a metal and be brown in color. When American nuclear physicists at the Manhattan Project succeeded, shortly before the end of World War II, in producing plutonium in sizable quantities in their nuclear reactors, they found that plutonium was indeed brown and a metal.

I previously posed the question about the measure of things. Why does a snow crystal look pretty much like any other snow crystal? We now know the answer: quantum mechanics. Quantum mechanics endows all things with stability. Life itself would be unthinkable without quantum mechanics. The stability of the genetic structure is based on quantum mechanics. The genetic information of a living organism is stored in the structure of its DNA, and DNA consists of molecules composed largely of electrons and the atomic nuclei of carbon, hydrogen, oxygen, and nitrogen. These molecules are stable because of the laws of quantum mechanics; they are as alike as two eggs. In essence, the laws of quantum mechanics are responsi-

3.4 Snow crystals display a remarkably high degree of symmetry, a consequence of quantum physics.

ble for the existence of life, for the fact that genetic information can be stored, for the resemblance of identical twins.

The uncertainty relationship shows that it is not possible to determine the location and velocity of an electron with absolute precision. One might therefore ask what it means to talk about Bohr's radius of the hydrogen atom. Why do so and yet say that uncertainty in fixing the location of the electron in a hydrogen atom is unavoidable? The answer to this question brings us directly to something that many physicists have been loath to accept: the probability aspect of quantum mechanics, which we are about to discuss.

CHAPTER 4

The Dice-Playing God
of Quantum Mechanics

> Everything existing in the universe is
> the product of chance and necessity.
> —DEMOCRITUS

/ N the summer of 1665 London and its environs were
ravaged by the plague. To halt the spread of the epidemic
on its campus, Cambridge University closed its doors and
sent its students home. Isaac Newton was among these
involuntary vacationers. He returned to his parental home
in Lincolnshire for what turned out to be a protracted stay
rather than a brief visit. The eighteen months Newton
spent at Lincolnshire were among the most productive of
his life. While there, he not only invented differential and
integral calculus—to the dismay of countless generations
of high-school students—but also laid the foundations of
classical mechanics, discovering the laws of mechanics
and gravitation. Up to the end of the nineteenth century,
physics progressed along the lines laid down by Newton;

it was marked by the elaboration, application, and perfection of Newton's ideas.

In a work entitled *Principia,* Newton presented the results of his research to the Royal Society of London, on April 28, 1686, a historic date. It signaled the end of an era, the end of the Middle Ages, the beginning of the Modern Age, the age of the natural sciences.

The mechanical systems described by Newton's laws are predictable exactly. Take a cannonball fired by a cannon. Once we know the exact speed with which the cannonball traverses the barrel we can figure out its trajectory. A branch of mechanics, ballistics, deals with problems of this type.

Edmund Halley, the English astronomer and mathematician, a contemporary and friend of Newton's, was interested primarily in comets. Comets have been the source of fear and superstition throughout human history. The flaring-up of a star with a long, luminous tail in its wake was seen as a threat to the harmony of the spheres, a danger to life. Comets were blamed for all sorts of disasters—wars, famines, epidemics. In 1578 a Protestant clergyman described a comet seen in the heavens as the "reeking smoke of human sin, full of horror."

It took a great deal of courage to assert that comets were normal celestial bodies subject to Newton's laws of celestial mechanics. Yet that is exactly what Halley did. In 1682 a comet visible to the naked eye dominated the night sky for a number of months. Halley measured its trajectory. He concluded that the comet was moving in an ellipsoid orbit around the sun and would take seventy-six years to complete its orbit. If that calculation was correct, this comet ought to be visible every seventy-six years. Halley consulted ancient astronomical charts and found that in 1607 and 1531 a big comet had indeed been sighted. Clearly it must have been the very same body he was seeing.

Halley figured out the exact day in 1758 on which "his" comet would presumably reappear. Neither Newton nor Halley lived to witness that momentous day, but at the precise date predicted by Halley, his comet appeared—a triumphant day for science in general and for Newton's classical mechanics in particular. It was now established that comets were simply normal heavenly bodies moving around the sun, that it was utter nonsense to link them to occurrences, catastrophic or otherwise, here on earth.

Halley's comet reappeared in 1834 and again in 1910, years in which nothing extraordinary happened on earth. The second appearance preceded the outbreak of World War I by four years. The next appearance of the comet is expected in 1986. There is no reason for panic; however, if 1986 should turn out to be a year of catastrophe it will certainly not be due to Edmund Halley's comet.

"Determinism" best describes classical or Newtonian physics: the laws of physics determine the behavior of physical systems. Everything is predetermined, nothing is left to chance, as exemplified by Halley's calculation of his comet's trajectory.

In the beginning of the nineteenth century a handful of scientists, among them the French physicist and mathematician Pierre Simon Laplace, dared to interpret the entire universe as one huge physical system governed by the laws of classical physics. The universe was seen as a sort of gigantic clockworks whose dynamics followed a predetermined pattern. In principle, Laplace concluded, the future of the universe was predictable, provided one knew the exact state of the universe at any given time in the past. There was no such thing as chance; everything was predetermined.

This idea gave rise to the question of free will and the role of God in the universe. When asked about the latter, Laplace replied that God was a hypothesis he could do without. I must confess that I felt very uneasy when, as a teenager, I first heard of Laplace's notion, for it leaves no room for free will. Everything is predetermined by the original state of the universe. Suppose you had breakfast at 7:35 this morning. If one accepts Laplace's implacable determinism, one would have to conclude that the hot gas cloud out of which our planetary system evolved billions of years ago was so designed that you had to sit down to your breakfast at 7:35 this morning, neither five minutes before nor five minutes later.

No one will deny that this conclusion is absurd, yet it is only an extreme example of the notion of determinism. If one truly believes that the laws of classical physics describe the universe unambiguously, one accepts the conclusion that the future is predetermined. One might even say that the future is contained within the present; there is no development. Today we know that Laplace's determinism is an unreasonable extrapolation. Classical physics, Newtonian physics, is an approximation that remains valid only if quantum phenomena are ignored.

The uncertainty relationship tells us that it is impossible to know simultaneously the exact location and velocity of an object. It is thus impossible to know the exact state of a physical system like the universe at a particular moment. Therefore it is also impossible to predict its future behavior with precision. Determinism has forfeited its claim to validity. The cold world of Newtonian determinist physics had left no room for cosmic growth, the future held nothing new in store.

Not so in quantum mechanics. The future does not appear unyieldingly certain; rather, it depends on various unpredictable chance occurrences. Quantum mechanics describes an open world, one that leaves room for development.

Laws of Probability

Let us now return to the question posed earlier. What does Bohr's radius signify if, according to the uncertainty principle, it is not possible to determine the precise location and velocity of an electron in a hydrogen atom?

Here quantum mechanics has a surprising answer. True, we cannot determine the exact location and velocity of that electron, but we can find the probability that it will

4.1 The space surrounding the hydrogen nucleus is divided into three spherical parts (see text).

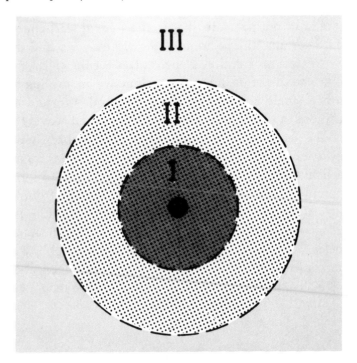

be located at a particular site. Let us assume that we are dividing the space around the nucleus of the hydrogen atom, that is, of the proton, into three parts (see figure 4.1).

The first part consists of all points whose distance from the nucleus is less than 1 Bohr (making that part a sphere); the second part consists of all points whose distance is more than one but less than 2 Bohrs. The third part encompasses all points whose distance from the nucleus is greater than 2 Bohr radii. Any physics student familiar with the mathematics of quantum mechanics can easily figure out the following probabilities:

Segment 1	32%
Segment 2	44%
Segment 3	24%
Total	100%

What do these figures mean? If we look at a hydrogen atom in its normal state, that is, the state of least energy, and determine the position of the electron, we find that the greatest probability, 44 percent, is for the electron to be in part 2. Let us, however, assume we are dealing not with a single atom but with a thousand. We find the positions of the electrons in these atoms and discover that about 320 are in part 1, about 440 in part 2, and 240 in part 3. We will not be able to make precise predictions for any particular atom; all we can do is establish probabilities. Undoubtedly electrons do exist whose distance from the nucleus is relatively great, greater than 2 Bohr radii. It is even possible for an electron to be very far from the nucleus, let us say more than 10 Bohr. The probability for this can also be figured out. It is very small indeed, about 0.0000005. One would have to examine more than a million hydrogen atoms to stand a chance of catching an electron attempting to "escape" from the nucleus. The structure thus is determined by positing a probability distribution. Quantum theory offers us an *exact* method for determining this probability distribution, which allows physicists to make precise predictions about probabilities.

We previously referred to the normal or ground state of the hydrogen atom—the state of least energy. What happens when we add energy to the atom, for example by irradiating it with light? In that event, the electron may suddenly move away from the nucleus. It is then said to leap to a higher state. Quantum mechanics makes very specific statements about these higher states. The uncertainty relationships tell us that the lowest states of the hydrogen atom are not the only ones to correspond to a well-defined level of energy; the higher ones, the so-called excited states of hydrogen, do also. The energy is said to be quantized—hence the term "quantum theory." Consequently, in the case of the hydrogen atom only very specific levels of energy are possible, a phenomenon referred to as the discrete spectrum of the hydrogen atom.

When a hydrogen atom is "excited," that is, when the electron moves away from the nucleus and possesses more energy than in the ground state, the electron will not necessarily remain in this higher, excited state for long. After a short time, the atom returns to its ground state, and the released energy is emitted in the form of electromagnetic radiation, such as light.

Every excited state of the hydrogen atom is described by a specific probability distribution for the electron. Figure 4.2 shows examples of these distributions. As we see, their structure can be quite complex, as complex as uncharted mountain terrain.

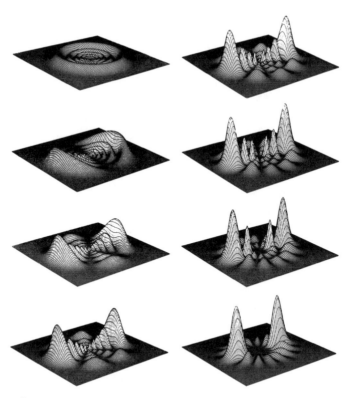

4.2 Some examples of quantum-mechanic distribution of electrons in the hydrogen atom, showing excited states of that atom. These distributions are, of course, three-dimensional, but given the limitations imposed by a two-dimensional page, this illustration shows only the distributions of the electron on the plane containing the atomic nucleus. For reasons of comprehensiveness the probabilities were multiplied by the square of the distance of the electron from the nucleus. (Reproduced by permission of D. Kleppner, Massachusetts Institute of Technology, Cambridge)

Excited atoms can be much larger than atoms in their normal state. For example, the typical size of the atoms in figure 4.2 is about 10^{-6} centimeters, more than a hundred times that of atoms in their normal state. Some highly excited atoms measure almost 1/100 (10^{-2}) millimeter

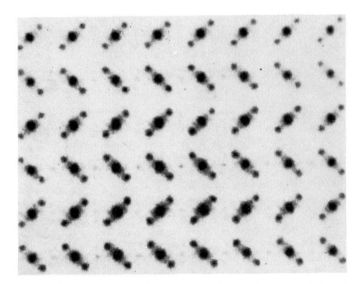

4.3 Pyrite molecules (FeS_2), shown 2.2×10^7 times their normal size. The large dark smudges are photographs of iron atoms (each of these atoms contains twenty-six electrons); the smaller, lighter areas are sulphur atoms containing sixteen electrons. (Reproduced by permission of M. J. Buerger, Massachusetts Institute of Technology, Cambridge)

(the size of some bacteria). With the help of special microscopes it is possible to "see" atoms (see figure 4.3).

The fact that quantum mechanics must restrict itself to probabilities imposes limits on our ability to predict the future. If, for example, we find in an experiment that the electron of a hydrogen atom at a given time is located at this or that place (let us say exactly 10^{-10} centimeters from the nucleus), we still do not know where it will be a little later.

In the past many physicists found it difficult to accept this limit on their cognitive powers, and some famous ones, Einstein among them, would not accept it at all. "God does not play dice," he used to say, rejecting the probability interpretation of quantum phenomena. He was wrong—God *does* play dice.

Today the probability interpretation of physical phenomena of quantum mechanics has found universal acceptance. This interpretation is nothing more than the consequence of the imperfection of our physical concepts. The essential concepts of physics—such as location, velocity, impulse—are defined within the framework of macroscopic physics, in the world of our daily experiences and of experiments. It is not surprising that we encounter problems in using these concepts to describe the phenomena of atomic physics or of elementary-particle physics. The probability interpretation of quantum mechanics is the compromise we have to make in order to describe the phenomena of atomic physics without relinquishing the customary concepts of the macroscopic world.

As far as quantum phenomena are concerned, the physicist wishing to investigate an atom is like a watchmaker who is asked to repair a wristwatch but finds that the only tools available are a hammer and chisel—crude implements considering the size of the object to be repaired. The watchmaker will have to use all the skill at his command to repair that little watch. The physicists likewise had to make some compromises in their effort to describe the world of the atom with the concepts at our disposal. The result is the quantum theory.

I never cease to marvel at how efficiently quantum theory can describe quantum phenomena. The whole business could have been far more complicated. All of quantum mechanics is governed by one parameter, Planck's constant $\hbar$, which, so to speak, is the gauge for the uncertainty that reigns in the atomic realm. Quite possibly, more than one parameter could have turned out to be important in the realm of atomic phenomena. In that event quantum mechanics, and with it atomic physics, would be significantly more complicated than it appears to be now. Fortunately that is not the case.

Another aspect of quantum mechanics also is astonishing. As we have seen, Newtonian classical physics is not valid in the realm of atomic phenomena. It is merely an approximation. The quantum theory does not derive from Newtonian physics. On the other hand, classical physics may be seen as a borderline case of quantum mechanics, the borderline one arrives at when one deals exclusively with phenomena for which Planck's constant is very small. This, of course, does not apply to describing the movement of an electron in a hydrogen atom, but it does apply to describing the movement of the moon around the earth.

One might think that even quantum theory is merely an approximation of the actual situation, that the time may come when we will have to scuttle that theory as well and introduce new methods, probably ones that contain quantum theory as an approximation. Many physicists used to believe that such a step would be required to explain the complex phenomena of elementary-particle physics. So far, these speculations have proved superfluous. Quantum mechanics has shown itself to be "unbeatable." It has demonstrated its validity not only for atomic physics but for a much broader area. It can even be used to describe the dynamic behavior of quarks, the smallest constituents of atomic nuclei. The validity of quantum mechanics is probably one of the most important lessons modern physics has taught us.

Neutron decay impressively confirms that quantum

theory can predict only probabilities and nothing more. We know that atomic nuclei are composed of protons and neutrons. They are generally stable; they do not change over time. Even individual protons are stable—but not isolated neutrons. Over time, a free neutron, one not part of the atomic nucleus, changes into other particles: it decays. This decay process gives rise to a proton, an electron, and another electrically neutral particle, a neutrino (see figure 4.4).

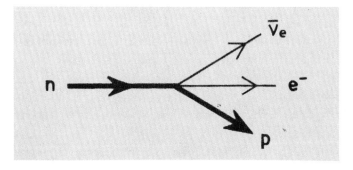

4.4 Neutron decay. A free neutron (n) is not a stable particle; it eventually decays into a proton (p), an electron (e^-), and a neutrino ($\bar{v}_e$).

We are unable to say exactly how long it takes a neutron to decay. Quantum theory tells us that all we can do is indicate the probability. There exists a 50 percent probability that neutrons decay within 11 minutes (this period —determined through experiments—is called the half-life). This means that out of a large number of neutrons, say 1000, half will no longer be "alive" after 11 minutes; after 11 minutes we will be left with only 500. If we wait another 11 minutes, only 250 will remain, and so on. The probability laws of quantum mechanics enable us to make predictions about many states, in this case about many neutrons. It is, however, not possible to make a firm prediction about a specific neutron. Of the 1000 neutrons in our experiment, it is likely that some will survive for as long as an hour. One might think that these long-lived neutrons will then soon die. Not so. Neutrons do not age. After an hour, the probability that these neutrons will decay in the following 11 minutes is no greater than before.

The example of neutron decay demonstrates the interplay of chance and necessity—an interplay of crucial importance for the structure of the universe and for the course of natural processes. There exists no law that says that a particular neutron *must* decay after a specific period of time. Even after the passage of a year, "our" neutron might still be alive, although the probability is exceedingly small.

If, however, we look at a great number of neutrons we can state that half of them will no longer be alive after 11 minutes. For a large number of neutrons there thus exists a deterministic law.

It may come as a surprise to learn that a neutron is unstable, whereas most atomic nuclei, which are after all composed of protons and neutrons, are stable. Nuclear physicists have a simple explanation for this phenomenon. When a neutron decays, a proton is formed. If the decaying neutron were inside an atomic nucleus the resultant proton would be there as well. It turns out that the protons and neutrons in the nucleus generally do not "like" the proton formed during neutron decay; they offer resistance to it. This resistance is so great that the proton cannot be produced at all, which means that neutron decay in the nucleus does not take place; the neutron inside the nucleus, then, is stable. We owe the existence of stable atomic nuclei to this circumstance.

In macroscopic physics, in our daily lives, the effects of

quantum-mechanical uncertainty are not felt. The fact that in the realm of the atom there is no such thing as absolute certainty, only probability, does not enter into macroscopic physics. Yet it is quite possible—for example, by using electronic amplifiers—to raise quantum-mechanical uncertainties to the macroscopic level. If we were to build a device that could register the spontaneous decay of a neutron and at the same time activate a relay, that relay could turn on a light switch or fire a gun—in short, it could set off a macroscopic process. The spontaneous decay of a single neutron would then have macroscopic effects.

Modern technology utilizes many quantum-mechanical phenomena. The fact that certain elements such as copper are good conductors of electricity while others are not ultimately is a phenomenon that can be understood only with the help of quantum theory. Some substances offer no resistance to electricity when cooled off—a quantum phenomenon called superconductivity that plays a major role in today's technology and whose future application seems very bright.

Quantum mechanics also explains many other familiar phenomena such as the color changes that occur in white-hot coal as it cools. We know that when a piece of coal is taken from a hot furnace it is white, but that gradually its color changes, slowly turning yellow and then red, until finally the coal ceases to emanate any visible light even though it may still be very hot. The color of the coal is somehow related to temperature. For a long time physicists were puzzled by this familiar phenomenon, but now quantum mechanics has enabled us to understand it. Many familiar properties of substances, such as the white sheen of silver, the yellow cast of gold, and the brown tinge of copper can also be understood with the help of quantum mechanics.

Quantum-mechanical fluctuations are responsible for biological mutations, purely accidental changes in the genetic structure. They are responsible for evolution in the organic world. Without mutations, without the constant and accidental changes in genetic structure, there would be no development, no life. Many other biological effects, such as aging and cancerous growths, are ultimately caused by quantum-mechanical fluctuations in the structure of cells.

The fact that natural processes do not follow a completely deterministic course, that cosmic development exists, is due to the quantum-mechanical uncertainty of the processes of nature. These uncertainties also imply that the future of the universe is not certain and that our fate in part depends on ourselves.

Newton and Laplace attempted to fit our world into the framework of the deterministic laws of physics. Quantum mechanics has released us from this straitjacket and given us an open world, one that has room for development and free will. Human freedom would be illusory without the uncertainty of the future laid down by quantum mechanics.

CHAPTER 5

Mysterious Fields

O N E of the first things we learn in life is connected with falling objects. When we let go of a stone in our hand, it will drop to the ground. One would think that nothing can be more obvious. And indeed, throughout human history this fact has become part and parcel of our consciousness. For millennia, hundreds of millions of people never thought of questioning why a stone should drop to earth when released. Most significant scientific discoveries have been made by people who speculated about phenomena that others accepted unquestioningly. Gravitation was one such case.

One day in the year 1666, young Isaac Newton (see figure 5.1) began to wonder why a stone falls to the ground. We can only speculate about what went through his mind, but possibly his thoughts went something like this:

"Wouldn't it be more natural for a stone simply to remain at rest where it is? I am standing here holding a stone in my hand. I can feel its pressure on my hand. It does not want to remain at rest where it is. Something

5.1 Isaac Newton (1643–1727). (Copperplate; Deutsches Museum, Munich)

---

compels it to relinquish its position of rest as soon as I let go. It falls to the ground. Suppose I were not here on earth but somewhere in the universe. I let go of the stone; nothing happens. The stone remains where it is. What else can it do? This means that the earth is responsible for the pressure on my hand. Yes, that's it—the earth attracts the stone. But there must be more to it than that. The difference between the earth and the stone is merely quantitative. Although the earth, unlike the stone in my hand, is vast, it still is nothing more than one great, big stone. The earth attracts my stone. However, the reverse must also be true, namely that the stone attracts the earth. Stone and earth attract each other. The earth is a big stone and it attracts my stone.

"Suppose I were holding two ordinary stones, one in each hand. Both are attracted by the earth, and both in turn attract the earth. But that makes sense only if both stones attract each other as well. All stones, all bodies, attract each other. I myself attract the stone, and the stone exerts attraction on me. All bodies attract each other, only we do not notice it because this attractive force is very weak. The earth is very big, and therefore its force of attraction is big enough for us to be aware of it. The earth attracts all bodies, including the moon revolving around it. The same force that compels the stone to fall to the ground also keeps the moon in its orbit around the earth. The moon 'falls' around the earth.

"The planets move around the sun. The sun is very massive. It attracts the planets. The planets attract the sun. The mutual mass attraction forces the planets to move around the sun. Without this attraction they would simply fly out into space and continue to move away from the sun. Yes, that's what it must be—mass attraction is a universal force that explains the movement of planets, the movement of the moon around the sun, and the apple dropping off the tree to the ground."

Newton's thoughts must have followed roughly along these lines when he made his famous discovery. That moment when he conceived the idea of the law of universal mass attraction, of gravitation, must count as one of the most significant ones of modern times. Only a little more than three hundred years has passed since that momentous discovery, but in this relatively short time our planet has changed more than in all the preceding eras.

The power of attraction between two bodies depends on their respective mass. The greater the mass, the stronger the power of attraction. But it also depends on the distance between the bodies. The power of attraction decreases in squared proportion to distance. If the distance between two bodies is doubled, the power of gravitation is only one-fourth as great as before—provided one ignores the volume of the body in relation to the distance (otherwise the situation becomes more complicated). If the distance is increased tenfold, the power of gravitation is only one-hundredth as great.

The development of physics and astronomy since Newton's day attests to the universal validity of his law. It describes the fall of a stone, the movement of the earth around the sun, the movement of the sun around the center of the Milky Way system, and the movement of galaxies billions of light-years away. It would be hard to find more persuasive proof of the universality of natural laws.

Why do two bodies attract each other? Newton was not yet in a position to answer this question. He was able to formulate the law of gravitation, but even though he tried, he was not able to discover a more profound reason for mass attraction. Newton believed the law of gravitation to be an action-at-a-distance law: the sun attracts the earth because its gravitational force acts directly on the earth, hurtling the distance between the sun and the earth, thus exerting remote attraction on the earth. Newton's conjecture was accepted uncritically by his successors until the early part of our century, when a minor official of the Swiss patent office at Berne by the name of Albert Einstein brought fresh insights to this problem.

Einstein posed the following question: suppose we

were to remove the sun from the universe, though of course that cannot be done. Still, suppose a magician were able to make the sun disappear. In science it is quite customary to engage in this kind of thought experiment.

Let us begin by asking Newton what would happen if the sun were suddenly to disappear. Newton would find the answer quite simple. If the sun suddenly disappeared so would its gravitational force. The movement of the earth would no longer be influenced by the sun's gravitational force, and the earth would simply continue to fly outward. The entire planetary system would cease to exist, and the planets would fly apart.

Einstein was not satisfied with Newton's answer, and this is why. In 1905 Einstein had published his theory of relativity, a theory that basically revised accepted notions of space and time. One aspect of this theory which has since been confirmed by numerous experiments concerns the velocity of the propagation of light. The speed of light is a constant of nature, almost exactly 300,000 kilometers/second. It furthermore follows from his theory that it is not possible to transmit signals at a speed exceeding the speed of light. Light takes about a second to travel from the earth to the moon. Einstein's theory established that it is not possible to send a signal to the moon in less than a second. The reason is easy to understand. In order to send a signal to the moon we must send out a radio light-signal in the direction of the moon. This signal travels with the speed of light, no more and no less.

Now we can also understand why Einstein was surprised by Newton's answer to the question about the disappearance of the sun. It takes light about eight minutes to get from the sun to the earth. If our magician suddenly were to make the sun disappear, we here on earth at first would notice nothing. According to Einstein's principle it would take at least eight minutes until the information "sun has disappeared" reached the earth. Its inhabitants could thus enjoy sunshine for a full eight minutes after the sun had ceased to exist. The catastrophe described by Newton would not occur until eight minutes later; then the world suddenly would turn dark, and the earth would fly off into interstellar space.

During the first eight minutes after the sun's disappearance the earth would continue to move in its habitual orbit as though nothing had happened. It could, however, do so only because of the persistence of some gravitational force after the sun's disappearance. Einstein's principle prevents our magician from making all gravitational force disappear instantaneously with the sun. Its gravitational force is eliminated only gradually, concurrent with the speed of light.

Based on this fact, Einstein concluded that the phenomenon of gravitation had nothing to do with any sort of remote action between material bodies, as Newton had assumed; rather, a local phenomenon had to be involved. In other words, the sun attracts the earth because the presence of the sun has changed the space in the vicinity of the earth. The sun is surrounded by a gravitational field in which the earth moves. The space between the sun and the earth is, so to speak, "filled" by this gravitational field. Since this involves a property of the space around the sun rather than a property of the sun itself, this field of force continues to exist for a time after the sun's disappearance. The elimination of the gravitational field takes place in the form of a shock wave, a gravitational wave that spreads spherically from the former site of the sun with the speed of light; it is similar to the shock wave produced when a stone is tossed into a pond.

Eight minutes after the disappearance of the sun that

wave arrives on earth, which then moves in a straight line. After some hours the gravitational wave reaches the outermost areas of the solar system, and after some years it arrives at the stars in the vicinity of the sun.

Gravitation has now introduced us to the concept of a field. As we shall see, it is a concept crucial to modern physics. For the time being, however, let us stay with gravitation. Einstein spent many years trying to understand the phenomenon of gravitation within the context of his general theory of relativity, which interweaves space, time, and gravitation in a single unit. None of these can exist independently. Gravitation turns out to be a property of space and time. A stone falls to the ground not because the earth attracts the stone directly, but because in the vicinity of the earth the properties of space and time undergo such changes that the stone has no choice but to fall to the ground: that is its natural movement. The gravity being exerted on the stone is a secondary phenomenon: this is the ultimate conclusion to be drawn from Einstein's theory of gravitation. Einstein's theory has many other interesting consequences, too, such as the deflection of light by the gravitational field of the sun or the existence of black holes.

Quantum Units of Space and Time

Before leaving gravitation I would like to take note of some quantum phenomena. The mass attraction of two bodies is determined by a fundamental constant, Newton's gravitational constant. (Physicists denote this constant with the letter G.) It is determined through experiments, and its value has been established with precision. We also know that this constant has the same value throughout the universe. Mass attraction, whether on earth or in remote galaxies, is governed by the same constant.

Physicists have not yet succeeded in marrying Einstein's theory of gravitation and quantum theory. Seemingly surmountable obstacles, not of a mathematical nature but connected with our traditional concepts of space and time, continue to crop up. The forces of mass attraction are closely linked to the structure of space and time. Therefore space and time would also have to be assigned quantum properties. Our conventional ideas of space and time are no longer unalterably applicable to very short distances or very brief time intervals. No one can predict the consequences this is bound to have. Some physicists believe, for example, that space and time have a foamlike structure. At any rate, we know exactly where our generally accepted notions of space and time must break down; our knowledge of the gravitational constant G, of Planck's constant h, and of the speed of light enable us to do so. Planck's units of space and time can be stated as follows: Planck's elementary length: 1.6×10^{-33} centimeters; Planck's elementary time: 5.4×10^{-44} seconds.

Planck's units of space and time are fundamental quantities in the natural sciences. It would certainly make sense to express all lengths and times in these units. Yet in this case that which makes sense is also very cumbersome. Instead of saying that there is a distance of 800 kilometers between Los Angeles and San Francisco, we would have to say that they are 5×10^{40} Planck lengths apart.

I recently chaired a conference in Sicily on problems of

gravitation. To announce a recess I told the assemblage that we were going to take a little break and asked them to return promptly in 2.2×10^{46} Planck time units. Exactly twenty minutes later the physicists returned to the meeting hall.

Astrophysicists estimate that the universe is about 20 billion years old. Expressed in Planck's units, we would say that the age of our cosmos equals 10^{61} Planck time units.

What is happening at distances close to Planck's elementary length or at times close to Planck's elementary time is one of the most interesting problems of modern science. Many theoretical physicists are working on it, but many years are likely to pass before we will have a fairly detailed picture of the structure of space and time.

Electric and Magnetic Forces

In addition to gravitation there exists another force all of us are familiar with. It can be observed in nature without the assistance of any complicated gadgets. That force is electricity. Two bodies that are electrically charged exert force on each other. If both carry identical charges, either positive or negative, they will repel each other; opposite charges attract each other. We have already seen that this basic characteristic of electrically charged objects is responsible for the structure of atoms and thus for the chemical properties of elements and substances.

Most objects in our daily lives carry no noticeable electric charge; yet while they are electrically neutral, they nonetheless are composed of electrically charged objects

—electrons and atomic nuclei, whose charges cancel each other.

The electric attraction of two dissimilarly charged bodies in some respects resembles the phenomenon of mass attraction. If the bodies are sufficiently far apart so that their volumes are small in comparison with their distance from each other, the gravitational force as well as the electric force decreases in proportion to their distance from each other squared. This analogy between electric and gravitational force does not, however, go very far. For example there is no such thing as gravitational repulsion of bodies, only attraction, but electric repulsion does exist.

Another important difference between mass attraction and electric force lies in the fact that the electric force is far more powerful than mass attraction. To illustrate this point let us look at the hydrogen atom. In the hydrogen atom the proton (atomic nucleus) and electron attract each other. As explained earlier, we are dealing with electric force. But beyond that the proton and electron also exert mass attraction on each other, in conformity with Newton's law. This force, however, is quantitatively far weaker than the electric force and therefore, as far as atomic physics is concerned, can be ignored.

In addition to electricity, the phenomenon of magnetism is also found in nature. We are familiar with the behavior of a compass needle in the magnetic field of the earth, or with the force that two iron magnets exert on each other.

As early as the beginning of the nineteenth century it was thought that the electric and magnetic forces were perhaps connected. In 1820 the Danish physicist Hans Christian Oersted accidentally made a startling discovery during a lecture. He found that magnetic forces can be produced when electric charges are moved to and fro. For example, a compass needle is deflected if an electric current passes in its vicinity. (Electric current passing through a metal wire is composed of numerous electrons that "flow" through the wire like water through a garden hose. An electric current thus consists of moving electric charges.) Oersted's discovery showed that there must be a close connection between electric and magnetic phenomena. Oersted himself introduced the now familiar concept of electromagnetism.

The phenomenon discovered by Oersted gave the English physicist Michael Faraday no rest. If the movement of electric charges was able to produce magnetic effects, then the opposite, namely the production of electric currents by a magnet, should also be possible. On August 29, 1831, after years of persistent experimentation, Faraday finally achieved the sought-after effect. He found that electric currents can be produced in the vicinity of an electric conductor (such as a copper wire) by moving a magnet to and fro. With this finding Faraday had discovered a very important principle; to this day we produce electric energy in our power plants by applying his discovery.

Faraday was born in 1791 near London, the son of a blacksmith. At age fourteen he began a seven-year apprenticeship in a bookbindery. He did not seem to have been very industrious at his job; he not only bound the books, but he also read them, particularly those dealing with the natural sciences. At age twenty-one he became the assistant of the chemist Humphry Davy at the Royal Institute of London, and there he stayed for the rest of his life. Upon Davy's death Faraday was appointed to his chair.

Faraday was a superb experimental physicist; however, he had no formal university training and his theoretical background, particularly in mathematics, was somewhat scanty. His intuition, on the other hand, was extraordinary, and this is probably why he was the first to develop the concept of electromagnetic fields.

In the beginning of the nineteenth century electric force as well as gravitational force was believed to be an action-at-a-distance force, not affected by the properties of the space between the two electrically charged bodies. Faraday dismissed this belief as nonsensical. He was convinced that there had to be something that connected two electrically charged bodies with each other. This simple assumption gave rise to the concept of electromagnetic fields. Faraday believed that the space between the charged bodies was, in a manner of speaking, filled with lines of force (see figure 5.2), and that although these lines of force emanated from the electrically charged bodies,

5.2 The electric-field lines created by the mutual attraction of two spheres carrying opposite electric charges.

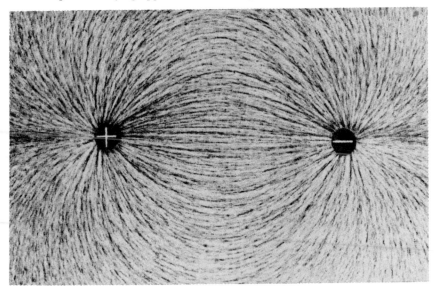

they led an independent existence and were an attribute of space. Electrically charged bodies attract or repel each other because the space between them has undergone a change—it is filled with electric lines of force.

The development of physics in the past hundred years has shown the concept of electromagnetic fields, as of all fields of force, to be of crucial importance. Modern theories of matter, which will be discussed later on, are field theories, that is to say, theories in which the concept of a field plays an essential part.

Faraday was not destined to construct the theory that describes the dynamical behavior of electromagnetic fields. That was left to the Scottish physicist and mathematician James Clerk Maxwell, who succeeded in translating Faraday's intuitive ideas into precise mathematical terms. In 1861 Maxwell formulated the remarkable equations of the electromagnetic field that bear his name. Since the formulation of Maxwell's theory of electromagnetism, two significant revolutions have taken place in the realm of physics—the quantum theory and relativity theory. Maxwell's theory has survived intact these two revolutions that have changed our basic ideas about the structure of space and time. His equations are as valid today as they were a hundred years ago. They describe the electromagnetic phenomena inside atoms and molecules, the electromagnetic processes of our technical devices and electronic instruments, and the electric and magnetic fields of galaxies, stars, and planets.

One interesting prediction of Maxwell's theory is the existence of free electromagnetic fields. Originally it was thought that an electric or a magnetic field had to be tied to an electrically charged body or a magnet. This idea, however, is no longer tenable. To see why let us conduct another thought experiment, similar to the earlier gravitation thought experiment.

In this experiment we are looking at an electrically charged sphere that is surrounded by electric lines of force. We once again ask our magician to make the charged sphere disappear. What happens to the electric field that surrounds the sphere? One might think that it would promptly disappear together with the sphere. However, this cannot happen, in keeping with Einstein's principle that no effect can spread more rapidly than the speed of light. In the moment the sphere disappears, someone at a distance of 300 kilometers could not yet have noticed its disappearance, for light would take a thousandth of a second to get to that person. Consequently the electric field at that point is as strong as it was prior to the disappearance of the sphere. Only after a thousandth of a second can the electric field change. Maxwell's theory offers a precise description of this change. After the disappearance of the sphere, an electromagnetic shock wave spreads from the former center of the sphere —almost like an explosion—and gradually engulfs the entire space. This eliminates the electric field (see figure 5.3). The sphere's field thus does not disappear all at once but gradually, and in the process an electromagnetic wave is formed. Maxwell's theory tells us that this wave always moves with the speed of light. Electromagnetic waves cannot stand still.

Maxwell predicted the existence of electromagnetic waves. In 1888 the German scientist Heinrich Hertz found proof of the existence of electromagnetic waves in nature. There is a direct line leading from Maxwell's and Hertz's discovery to the present age of electronics.

When Maxwell took a closer look at the implications of

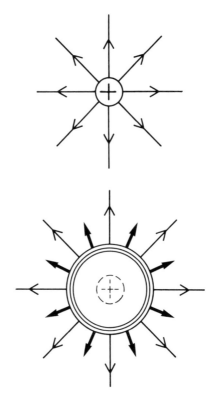

5.3 An electrically charged sphere suddenly disappears, whereupon an electromagnetic wave spreads from the center of the electric field with the speed of light. The electric field gradually is extinguished.

his theory, he himself was surprised to discover that electromagnetic waves always travel with the speed of light. Following the ideas of Faraday, Maxwell set up a daring hypothesis, to wit, that light was nothing more than an electromagnetic phenomenon. He was right. The light waves registered by our eyes are electromagnetic waves. The interplay of light and matter, for example the reflection of light on a mirror, is an electromagnetic phenomenon. Light waves fall on the silvery metal surface of the mirror, interact with the electrons of the metal, and are in turn reflected by it.

How does an electromagnetic wave travel through space and in what medium? We tend to connect the concept of *wave* with the image of a stone tossed into a calm lake, with concentric waves forming and spreading over the water's surface. The water here serves as the medium on which the waves spread. Does such a medium exist in the case of electromagnetic waves? The answer to this is astonishing yet simple: there is no material medium in which electromagnetic waves spread. An electromagnetic field, like a gravitational field, is nothing more than an intrinsic characteristic of space and time.

Regions containing electromagnetic or gravitational fields might be considered to have special distinguishing features. Something is happening in them; physical processes are taking place. Regions that do not contain such fields are calm. Nothing is more boring than empty space.

Let it be noted that the cosmos contains no space that is truly boring. Even intergalactic space far removed from any galaxy is filled with electromagnetic fields, with waves rushing through space hither and yon with the speed of light.

I mentioned earlier that Maxwell's theory survived the quantum revolution in physics intact. That is true, but with one important limitation. Maxwell's theory has to be interpreted properly if we want to avoid coming into conflict with the quantum theory; Einstein pointed this out, in 1905. He found that Maxwell's theory is compatible with Planck's quantum theory only when the energy of an electromagnetic wave, particularly a light wave, is transferred in small, well-defined quantities. These light "quanta" later were given a special name—they are called *photons,* light "particles."

The amount of energy carried by a light quantum depends on the wavelength of the light. Blue light has a shorter wavelength than red light. Consequently, blue light photons carry more energy than those of red light. Roentgen waves are electromagnetic waves much smaller in length than those of visible light. The photons of Roentgen waves (X rays) consequently are more energetic than those of ordinary light, and thus capable of penetrating the human body.

Quantum theory also interprets the attraction and repulsion of electrically charged bodies differently than did classical physics before Planck and Einstein. Quantum theory holds that electric and magnetic fields, like particles such as electrons or protons, also possess quantum properties and are subject to uncertainty relationships.

Quantum mechanics describes the attraction of electrically charged bodies in a seemingly curious manner—as the exchange of photons. Photons continuously wander about between the charged objects, and this wandering produces a force: electricity (see figure 5.4).

With his light-quantum hypothesis, Einstein introduced a new particle into physics, the photon. This particle differs from those we were introduced to earlier, the proton and electron, by its mass. Both the electron and the proton possess mass. The mass of the proton, for example, is 1.7×10^{-24} grams. The photon has no mass: it is a massless particle.

Yet the photon does have a given amount of energy that, like a busy insect, it constantly carries from place to

5.4 Electric forces such as the attraction of objects with opposite charges shown here are produced by an exchange of photon quanta (here represented by the ball).

---

place. We may find it difficult to imagine particles that have no mass but are able to transfer energy, since we tend to link the transfer of energy to the notion that energy is something contained in a massive body—like the kinetic energy of a moving vehicle. Nonetheless, such is the case.

Photons also are unusual in yet another respect: a photon can never be at rest. Traveling with the speed of light, it is constantly on the move. That is the conclusion drawn from the fact that light waves always move with the speed of light. Photons, as "particles" of light, thus must act in like manner.

Despite these specific characteristics the photon is a particle similar to other particles like electrons, protons, and neutrons. We will discuss these particles in the following chapter.

CHAPTER 6

Matter and Antimatter

/ N the lobby of the European Nuclear Research Center, CERN, at Geneva stands a strange apparatus, an unwieldy box that produces a loud clatter when activated by a button located on its sides. This device, called a spark chamber, records cosmic rays. The earth is constantly being bombarded by particle radiation from space. We still know very little about the sources of these rays; perhaps they originate in the center of our galaxy.

These cosmic rays generally interact with the atomic nuclei of the gas atoms in the upper layers of the atmosphere. In the process new particles form that are propelled toward the earth. It is these particles that are recorded by CERN's spark chamber. They rush through the chamber with lightning speed, producing sparks visible to the naked eye and an audible clatter.

Like CERN's spark chamber our body also is constantly subjected to cosmic radiation. We are continuously being penetrated by cosmic ray particles. Frequently these particles collide with one of the atomic nuclei in our bodies and destroy it. This nuclear reaction generally takes place unknown to us; after all, one atomic nucleus more or less in our body is no big deal.

In the early 1930s Robert Millikan, working at the California Institute of Technology, undertook a closer study of cosmic rays. Working with him was Carl Anderson, the son of Swedish immigrants. To help him in his research, Anderson constructed a large cloud chamber, a device allowing him to plot the tracks of cosmic ray particles. When a particle passes through this chamber it leaves a trail somewhat like the vapor trail of a jetliner. By producing a magnetic field inside the cloud chamber, one can deflect electrically charged particles. The particle paths appear to be curved. The degree of their curvature contains clues to the velocity and electric charge of the particles.

When Anderson installed his new cloud chamber he examined a series of particle tracks. He was pleased to find that the observed particles were familiar objects, namely, positively charged protons and negatively charged electrons. On August 2, 1932, Anderson went about his daily task of photographing particle tracks in his cloud chamber. When he studied the tracks after developing his photographs he noticed something peculiar: a track that at first glance looked like that of an electron had the wrong curvature. The particle behaved like an electron with an erroneous, positive electric charge. After examining some more pictures Anderson discovered additional candidates for positively charged electrons.

On first seeing Anderson's results Millikan reacted with skepticism. He tried to persuade Anderson that his allegedly new, positively charged particles were simply ordinary protons. But before long Millikan, too, became convinced that Anderson had discovered a new particle. The mass of this particle was the same as that of an electron, and it carried a positive charge. Anderson called his particle the *positron*. It soon emerged that Anderson had discovered not just another particle, but a very special one at that. As we shall see, the positron is closely related to the electron. For his discovery Anderson was awarded the Nobel Prize for Physics in 1936.

Toward the end of the twenties, a young theoretical physicist in Cambridge, England, Paul A. M. Dirac, attempted to apply the then new quantum theory to Einstein's theory of relativity. In 1928, Dirac succeeded in setting up an interesting new equation, now known as Dirac's equation to physicists the world over. (Dirac is now in his eighties and teaches at the University of Tallahassee, Florida.) It describes in simple terms the movement of electrons in atoms. One of the important products of Dirac's theory was the discovery that his equation not only accurately describes the behavior of electrons in atoms but also predicts the existence of particles of the same mass as electrons, though with the opposite charge. Dirac called the new, positively charged particles *antiparticles* of the electrons.

When Anderson made his discovery he had no inkling of the theoretical work being done at Cambridge. Not until 1933 did it become clear that Anderson had in fact found the particles predicted by Dirac. The positron proved to be the antiparticle of the electron. Its symbol is e^+.

Anderson's discovery was the first step into a new world, the world of antimatter. Every particle has its own antiparticle. Thus there exists an antiparticle of the proton, the antiproton ($\bar{p}$). The antiproton has the same mass as the proton but carries a negative charge. It was discovered in 1955, not in cosmic rays but with the help of particle accelerators. In the early fifties an accelerator was built at the University of California at Berkeley that could bombard matter with protons at high speeds. In the course of these reactions new particles were produced. In October 1955, physicists at Berkeley announced that they had observed a negatively charged particle of the same mass as the proton. The antiproton had been discovered.

How do matters stand with regard to the photon, the particle of light? Do antiparticles of the photon also exist? The answer to this is yes; it is the photon itself. The photon is an exception in that it is its own antiparticle.

Dirac's theory predicts a rather strange fact, one that crops up in science fiction in startling ways. When a particle and an antiparticle collide, an explosion occurs—the two particles annihilate each other and are transformed into pure energy. To understand the process we must again fall back on an important aspect of Einstein's theory of relativity, the equivalence of mass and energy.

The concept of mass is well known. A liter of water has a mass of 1000 grams, whereas a proton has a mass of only about 10^{-24} grams. Einstein discovered that every mass has to be assigned a specific energy, the amount expressed in his famous formula $E = mc^2$ (energy equals mass multiplied by the square of the velocity of light). According to this relationship it is possible to convert mass into energy, and vice versa. The mass of an object might be considered as frozen energy.

Mass into Energy

What does such a conversion of mass into energy look like? Let us take the simplest example, the annihilation of electrons and positrons. When Anderson examined the tracks of his positrons in his cloud chamber he found that some of them broke off abruptly, as though they had suddenly disappeared from the world. And that was indeed what had happened. Those positrons had accidentally met up with some of the electrons in the gas atoms, an encounter in which the two particles destroyed each other.

Let us look at this process more closely and assume we have an electron and a positron at a distance from each other of, let us say, 1 centimeter. Since the two particles carry opposite charges they attract each other and move toward each other. Suddenly, both disappear, and their place is taken by two photons rushing from the site of the encounter with the speed of light (see figure 6.1).

The energy of these photons can be measured with relative ease. The result shows that the sum of the two

---

6.1 An electron and positron annihilate each other, producing two photons in the process. The energy of the two inflowing particles is recovered in the energy of the two outflowing photons (a). An electron-positron pair is created out of nothing by the collision of two photons. This process can only happen if the photons possess the energy required for the production of the two massive particles (b).

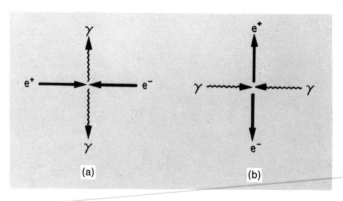

photon energies is equal to the energy assigned to the two particles, the electron and positron, on the basis of Einstein's equation. The sum of the mass of the electron and the positron has been converted into energy, in this case the energy of the two photons. One might also say that matter has been converted into light. (In this sense mass might also be called "frozen light.") The photons produced by the electron-positron annihilation, however, have far greater energy than the photons of visible light. These high-energy photons are called gamma rays or gamma quants (abbreviated as γ quant).

The reverse process, the transformation of energy into matter, can also be observed. When two photons approach each other they can suddenly change into an electron and a positron, into an electron-positron pair (see figure 6.1*b*). In this manner new particles can be born out of "nothing," out of pure energy. This process, however, can take place only when the photons in question possess enough energy to produce an electron-positron pair. The total energy must be conserved, even in such an exotic process as discussed here. After the creation of the pair the original energy must remain the same. Although new particles can be produced out of "nothing," energy cannot: energy is always conserved.

In line with Einstein's equivalence of mass and energy, it has become customary in physics, particularly in elementary-particle physics, to express the mass of a particle not in grams or other units of mass but in units of energy. The electron volt is such a unit of energy. One electron volt (eV) is that infinitesimal amount of energy an electron absorbs when it passes through the voltage gradient of a volt. The mass of an electron thus is equivalent to 0.5 million eV, or 0.5 megaelectron volts (MeV). In subsequent chapters I will occasionally refer to the GeV (gigaelectron volt). One GeV is equivalent to 10^9eV, or a billion eV.

It is very simple to determine the energy of the photons produced by an electron-positron annihilation. If both particles are at rest prior to the destruction, the energy of each newborn photon is given by the electron mass: 0.5 MeV.

Let us now turn to the second electrically charged particle we already know, the proton. A proton is almost two thousand times as heavy as an electron. It has a mass of 938 MeV, almost 1 GeV. An antiproton has, of course, the same mass.

Let us follow the same procedure with the proton-antiproton system as with the electron-positron system and assume that we are bringing a proton and an antiproton close to each other. In that event we will possibly expect the same thing to happen as in the electron-positron annihilation: the proton and its antiparticle destroy each other, resulting in two photons that carry off the available energy. The energy of these photons naturally must be correspondingly larger. Each of the two γ-quants carries the considerable energy of 938 MeV.

Yet when we conduct this experiment in the laboratory, we find to our surprise that the annihilation of the proton and antiproton into two photons occurs only very rarely. What we normally get is something entirely different, namely the production of a whole series of particles comprised largely of new particles called mesons, which we will discuss shortly, and, frequently, γ-quants as well (see figure 6.2). We further find that the number of particles produced varies considerably. At times their number is relatively small, perhaps four, and at other times it is

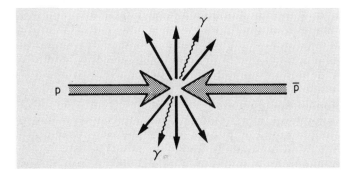

6.2 A proton and an antiproton annihilate each other. In contrast to the electron-positron annihilation, this process gives rise to new particles, including gamma quanta.

substantially larger. The number of particles produced depends on the energy of the original proton and antiproton. If both have a high velocity, meaning high energy, prior to collision, they may give rise to a great number of new particles.

In 1981 an accelerator able simultaneously to accelerate protons and antiprotons to enormously high energies and to hurl them at each other was put into operation at CERN (figure 6.3). The energy of these particles equals approximately 270 GeV, or almost three hundred times the energy of a resting proton and antiproton. Their speed before collision is practically the same as the speed of light, namely 0.999994 the speed of light. The collision of the two particles releases energy of twice 270 GeV, or a total of 540 GeV. This energy would suffice to produce 287 proton-antiproton pairs, or more than half a million electron-positron pairs, out of the "vacuum." Although these specific processes are possible in principle, they do not take place in practice. Still, at the proton-antiproton

6.3 Aerial view of the Geneva area with the European Nuclear Research Center, CERN, in the foreground. The large ring-shaped proton-antiproton accelerator is located in an underground tunnel to the left of the CERN complex and therefore cannot be seen here.

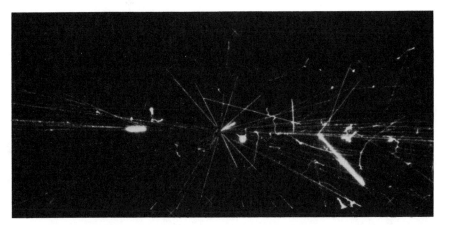

6.4 A picture of a proton-antiproton collision taken by the research team UA5. The frontal collision of a proton with an energy of 270 GeV and an antiproton of like energy gives rise to the production of numerous new particles out of the energy present in the collision. (Reproduced by permission of the UA5 Collaboration, CERN)

collisions at CERN, a substantial number of new particles are typically produced. Figure 6.4 illustrates one such process (see also figure 6.5).

Mesons

What particles are produced in a proton-antiproton collision? On closer examination we find a new, positively charged particle, one whose existence has not previously been mentioned, possessing a mass of about 140 MeV. It is the π^+ meson, discovered in the late 1940s in cosmic rays. This π^+ particle also has its antiparticle carrying the corresponding negative charge; it is the π^- meson.

Upon still closer examination of the proton-antiproton annihilation we find that there exists still another particle

6.5 The huge particle detector of the UA1 Collaboration proton-anti-proton accelerator at CERN (Photo: CERN). This instrument, the product of the joint efforts of more than 100 physicists from eleven institutes in Europe and the USA, made possible the detection in January 1983 of W bosons, the particles that transmit weak interaction. The name UA1 stands for Underground Area One. (Reproduced by permission of the UA1 Collaboration)

whose mass is only slightly less than that of the π^+, specifically, 135 MeV. This new particle is neutral. It is called the neutral π meson (π^0). Like the photon, the π^0 meson is its own antiparticle.

At this point you may perhaps ask why the π meson remained unknown for so long. The proton, after all, has been known for some time. Why, then, not the π meson? The answer is quite simple. The π meson is not a stable particle like the proton. It decays. We have already made the acquaintance of another unstable particle, the neutron, which decays within about 11 minutes. The π meson also decays, and it does so after a very brief time, practically at birth. The charged π meson decays within about 10^{-8} seconds. In elementary-particle physics, 10^{-8} seconds is a rather long time. In that span the π meson can cover a considerable distance, a few meters and more, before giving up its life. Upon examining the decay products of charged π mesons we find that ultimately they invariably decay into an electron and/or a positron and into a specific number of neutrinos, those neutral particles we came across in the decaying neutron. The π^+ meson always decays into a positron plus neutrinos, the π^- meson into an electron plus neutrinos. The electric charge of the mesons finds itself in the electric charge of the electron and/or positron. This must happen, for the electric charge, like energy, remains constant in all processes of nature. Charge can be neither produced nor destroyed. Two particularly simple decay processes of the charged mesons (although not the most important ones) are

$$\pi^+ \to e^+ \nu_e$$

and

$$\pi^- \to e^- \bar{\nu}_e$$

(ν_e: neutrino; $\bar{\nu}_e$: antineutrino).

The neutral π meson makes it very easy for itself. It decays into two photons, and does so after the very brief period of 10^{-16} seconds. This is so short a time that in the laboratory the neutral π mesons cannot be observed easily. They can be recognized only by their decay products, the photons. The photons observed in the proton-antiproton annihilation, for example, generally are the decay product of a π^0 meson produced in the course of the $p - \bar{p}$ annihilation that promptly decayed.

The decay of the neutral π meson is reminiscent of the electron-positron annihilation. It may seem strange that a neutral π meson should decay into two photons, just like the electron-positron system. But as we shall see, that is no accident, but, in fact, directly related to the internal structure of the π^0 meson.

Let us take a brief look at two other particles, the neutron and the neutrino. In my discussion of quantum-mechanical probability I referred to the fact that the neutron is not stable, but disintegrates into a proton, electron, and neutrino. As far as energy is concerned, this decay is entirely legitimate, for the neutron has a mass of 939.57 MeV, and the proton one of 938.30 MeV. The neutron thus is 1.27 MeV heavier than the proton. What is striking here is not so much the fact that the neutron is somewhat heavier than the proton, but that their masses are almost identical. The proton carries an electric charge; the neutron is neutral. The two particles thus differ completely with regard to their electromagnetic properties. Why, then, is their mass practically identical? Just think: the proton and the positron carry the same electric charge yet their mass is very different. What is the reason for the near equivalence of the masses of protons and neutrons?

The first scientist to wonder about this phenomenon

was Werner Heisenberg, who began to devote himself to this problem in the early 1930s, right after the discovery of the neutron. Heisenberg deduced a series of consequences for the dynamics of atomic nuclei from the approximate equivalence of the mass of protons and neutrons, particularly with respect to nuclear structure. Physicists call this the isospin symmetry. The reason for the approximate equality of the respective masses was not known until recently; only during the past fifteen years has light been shed on this matter. But let us leave this discussion until the next chapter and instead devote a little more time to the neutrino.

Neutrinos

The neutrino is probably the most peculiar elementary particle known to us. In chapter 14 we will see that the neutrino may well be the most important of all elementary particles.

The neutrino was not discovered until fairly recently, in the 1950s. Like the neutron, it is electrically neutral, hence its name, coined by Enrico Fermi. Neutrinos have a very remarkable characteristic: their interaction with normal matter is extremely weak. They can easily pass through substantial accumulations of matter, such as the earth or the sun. In only rare instances will they interact with matter. If the sun were to have a radius of a light-year, a neutrino beam would still pass through it almost unimpeded.

Various research centers, CERN among them, have constructed special devices for the production of neutrino beams. CERN's neutrino beam is produced on its home territory, where it proceeds through various experimental devices; only very few of the neutrinos cause reactions, and these are closely studied. Most of the neutrinos, however, simply continue on their way unperturbed, leaving the CERN site, passing through the nearby Jura mountain range, and out into space. Nobody can stop them. CERN's neutrino beam will continue to wander through the cosmos billions of years hence, uninhibited by either stars or gas clouds.

Neutrinos possess still another unique characteristic: they are very light. We do not know whether they possess any mass at all. It is quite possible that they have none, like photons. Still, many physicists are convinced that they do possess some mass, even if only an infinitesimal amount. In 1979, physicists at the ITEP research institute of the Academy of Sciences at Moscow claimed to have found proof that neutrinos possess a mass of about 20 eV. To date, this finding has not been corroborated by any other research center, and it most likely will be some time before we will know unequivocally whether or not neutrinos possess mass. But we do know that their mass cannot be very great, at most about 30 eV. At any rate, neutrinos are very light particles, more than ten thousand times lighter than electrons.

With this brief discourse on neutrinos we conclude our discussion of particles found in the cosmos: of protons and neutrons, the building blocks of atomic nuclei, and of electrons, the building blocks of atomic shells. Other important particles are photons, mesons, and neutrinos. All these, including the neutrinos, have antiparticles. The charges and mass of particles are given in the table below.

Matter and Antimatter

Particles of Matter

Particle	Charge	Mass (in MeV)
Proton	+1	938.30
Neutron	0	939.57
Electron	−1	0.51
π^+ Meson	+1	139.57
π^0 Meson	0	134.96
Photon	0	0
Neutrino	0	?

The corresponding antiparticles have the same respective mass and opposite charge. The photon and π^0 meson are identical with their corresponding antiparticles.

Why Atomic Nuclei Are Stable

Protons and neutrons are the building blocks of atomic nuclei. Complex atomic nuclei, such as the nucleus of the uranium atom, are composed of hundreds of protons and neutrons. What causes protons and neutrons to combine into atomic nuclei? An answer to this question does not require a study of complex atomic nuclei; we merely need to study the behavior of two protons. Suppose we are looking at two protons 1 centimeter apart. Since both carry a like charge they repel each other. Now we force them closer to each other. Nothing much happens, except that the electric force of repulsion gradually increases. The protons move closer to each other until they are only 10^{-12} centimeter apart. The electric force of repulsion has meanwhile grown stronger. Suddenly we notice that the two protons no longer repel each other, but that, on the contrary, they attract each other strongly. At a distance of 10^{-13} centimeters this new force is about a hundred times as powerful as the electric force.

We have just learned about an important new phenomenon, the so-called *strong interaction*, which occasionally is also called the strong force or the nuclear force. It was not until the 1930s that physicists realized that this new force existed alongside electromagnetic force. Nuclear force has the peculiar characteristic of being detectable only at very small distances, somewhere in the neighborhood of 10^{-13} centimeters. At longer distances it promptly vanishes. That explains why nuclear force was discovered relatively late, much later than electric force, which was known back in antiquity.

Nuclear force acts not only between protons, but also between neutrons or between neutrons and protons. If a proton and a neutron are brought together, the two particles attract each other and form a bound state, the deuteron, an object consisting of a proton and a neutron.

Nuclear force plays a vital role in the structure of our world. An atomic nucleus contains protons and neutrons. The protons repel each other electrically. If the nuclear force were switched off suddenly, global catastrophe would result; all atomic nuclei would promptly explode.

Not all the particles examined by us thus far take part in the strong interaction: only protons, neutrons, and π mesons. All other particles (electrons, neutrinos, photons) ignore this force. Electrons or neutrinos can penetrate an atomic nucleus without any difficulty. Their path is not affected by the strong force.

What is the deeper reason for the existence of the strong nuclear force? In the next chapter we will find out that it is related to the structure of the strongly interacting particles. The protons, neutrons, and π mesons prove to be compounds, which are composed of still smaller constituents, quarks.

In closing this chapter I would like to call attention to a peculiar circumstance to which we will return later.

Every particle mentioned has its antiparticle. These particles and antiparticles generally behave in like fashion. Physicists even speak of a particle-antiparticle symmetry. We can utilize the antiproton, antineutron, and positron to produce antimatter: antiatoms, antimolecules, and so forth. An antihydrogen atom, composed of an antiproton and a positron, behaves exactly like a normal hydrogen atom. When a fragment of antimatter is brought together with matter, however, all hell breaks loose. Matter and antimatter destroy each other, producing in particular large numbers of photons. The amount of energy released in the process is enormous. The energy released in the annihilation of a few kilograms of antimatter and a corresponding amount of matter would be sufficient to meet the energy needs of a state like California for a whole year.

The symmetry of matter and antimatter presents physicists with a puzzle. Matter, of which we, the earth, and the sun are made up, is composed of protons, neutrons, and electrons. There is nothing to indicate that a large accumulation of antimatter exists anywhere in our solar system. The same holds true for our galaxy. If antimatter were to exist anywhere in our Milky Way system—for example, in a star composed of antiparticles—we would undoubtedly see annihilation processes, since contact between antimatter and normal matter would be unavoidable. It would result in the radiation of a vast number of γ quants. Scientists have searched for such γ rays without success. For that reason it appears certain that our galaxy contains no antimatter. That same certainty does not, however, hold true for other galaxies. We cannot rule out the possibility that somewhere in the universe there are galaxies composed entirely of antimatter. On the basis of observation we cannot say which galaxies consist of matter and which of antimatter; an antigalaxy would look exactly like a normal galaxy. However, there are reasons to believe (we shall come back to this later) that our cosmos contains practically no antimatter at all. As we shall see, this peculiar asymmetry between matter and antimatter is directly connected with processes that took place about twenty billion years ago, soon after the birth of our cosmos.

The recent discoveries in elementary-particle physics were made almost exclusively in experiments using accelerators. In the accelerator laboratories, protons or electrons are accelerated to very high energies with the help of electromagnetic fields. After the acceleration, the particles move practically with the speed of light.

There are two methods of acceleration. One involves furnishing the electrons or protons with energy along a straight-line acceleration track called a linear accelerator. The other method involves letting the particles repeatedly run around circular tracks. With each revolution the particle acquires more energy. In this way high energies can be produced rather easily.

The accelerated particles are either directed toward a piece of matter, the target, or brought into collision with other accelerated particles. We previously mentioned this latter possibility in connection with the proton-antiproton annihilation. The collisions of particles are studied by elementary-particle physicists, and the phenomena observed allow us to draw conclusions about the structure of matter at very small distances.

During the past twenty years, experimental research in the field of particle physics has become very costly. For that reason physicists increasingly have been conducting their experiments in large, often international, research centers. At the present time, the most renowned such institutes are:

The European Nuclear Research Center (CERN) at Geneva, Switzerland (figure 6.3, p. 00)

The Fermi National Accelerator Laboratory (FNAL) near Chicago, Illinois (figure 6.6)

The Stanford Linear Accelerator Center (SLAC) at Palo Alto, California (figure 6.7)

The German Research Center (DESY) at Hamburg

The experimental aspects of the findings discussed in the following chapter were for the most part made in these institutions.

6.6 An aerial view of the Fermi National Accelerator Laboratory (FNAL or Fermilab) in Batavia, Illinois. The largest circle is the main accelerator. Three experimental lines extend at a tangent from the accelerator. The sixteen-story twin-towered central laboratory is seen at the base of the experimental lines.

6.7 The Stanford Linear Accelerator Center (SLAC) near Palo Alto, California. The long, linear tunnel that houses the vaccum pipe in which the electrons are accelerated crosses the freeway from San Francisco to San José.

CHAPTER 7

Quarks: The Stuff
of Matter

"Three quarks for Muster Mark"
—JAMES JOYCE
Finnegans Wake

H O W does one find out whether or not an object is composed of yet smaller objects? By looking for them. Ernest Rutherford used alpha rays to look into atoms (alpha particles are simply the atomic nuclei of helium atoms). A radiologist uses X rays, that is to say, photons, to look at a patient's lungs. We want to examine the interior of the proton, and so we have to build a huge X-ray machine, one able to penetrate distances of less than 10^{-13} centimeters. Beams of electrons are the suitable instruments for measuring the substructure of a proton. Electrons do not interact via strong interactions, and therefore can penetrate a proton without difficulty.

In early 1957, Wolfgang Panofsky of Stanford University, California, submitted a proposal to the Atomic Energy Commission calling for the construction of a gigantic electron accelerator, a linear model able to accelerate electrons to an energy of more than 20,000 MeV. In May 1959, Congress approved President Eisenhower's request for the $100 million project. Seven years later, on May 21, 1966, scientists began working with the new accelerator, SLAC (Stanford Linear Accelerator Center), at Palo Alto (figure 7.1). The experiments which began there turned out to be of vital significance.

The SLAC accelerator made it possible to guide high-energy electrons toward atomic nuclei, that is, toward protons and neutrons. This experiment was similar to Rutherford's, but it involved electrons rather than α-particles, and atomic nuclei rather than atoms.

If a proton were composed of still smaller objects one might expect that the electron penetrating it would occa-

---

7.1 View of the accelerator track of the Stanford Linear Accelerator Center (SLAC) at Palo Alto. Shown here is the vacuum tube in which electrons are accelerated to an energy of about 20 GeV.

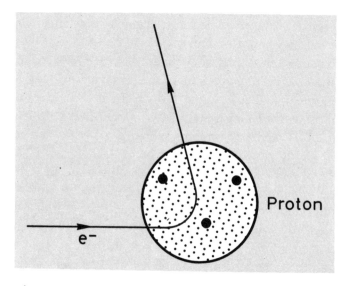

Proton

e⁻

7.2 An energetic electron penetrating a proton is deflected in its flight by one of the proton's constituents. The force responsible for this phenomenon is the electric attraction and/or repulsion operating between the electron and the constituent. Of course the only objects that can be observed in this process are the electrically charged constituents of the proton. The electron ignores electrically neutral building blocks of the proton.

sionally collide frontally with one of the constituents and be deflected by it (figure 7.2). And, indeed, the Stanford physicists did find that electrons traveling near the speed of light tended to change their direction abruptly when colliding with matter (such as, for example, a chunk of iron; see figure 7.3), thereby proving that protons and neutrons are composed of smaller components. Upon closer investigation it was found that protons are composed of three different building blocks.

This discovery by the SLAC scientists proved a hypothesis advanced in 1964 by two American physicists, Murray Gell-Mann and George Zweig. They had

7.3 The electron detector used by a combined research team of SLAC and MIT scientists to examine the detailed structure of protons at SLAC. (Photo: SLAC)

hypothesized that protons and neutrons were composed of three building blocks, which Murray Gell-Mann, in a tribute to the writer James Joyce, dubbed "quarks."*

In 1970 similar experiments were begun at the CERN research center. However, in their assault on the proton, the CERN scientists used neutrinos rather than electrons. The results obtained at CERN also supported the quark hypothesis (see figure 7.4).

But enough history! Let's take a look at what we have since learned of the structure of the proton. It takes two

*For a more detailed description of quark theory see Harald Fritzsch, *Quarks* (Basic Books, 1983).

7.4 View of the large CDHS detector at CERN. Its name is an acronym of the institutes involved in its construction and experimental application: CERN, Dortmund University, Heidelberg University, and Saclay Research Center, Paris. This detector allowed us to explore the interior of the protons with the neutrino beams produced at CERN. The results corresponded to the predictions of the quark model.

different quarks to describe the stable nuclear matter observed in the universe (figure 7.5), which we name u and d. This notation is easy to understand. Let us write two quarks vertically, one above the other:

$$\text{Quarks} \sim \begin{pmatrix} u \\ d \end{pmatrix}$$

The letters u and d stand for up and down. A proton is

7.5 The fine structure of matter. Atoms are composed of electrons and a nucleus. The nucleus is composed of protons and neutrons. Three quarks are the building blocks of protons and neutrons.

---

composed of two *u* quarks and one *d* quark (see figure 7.6), and a neutron of two *d* quarks and one *u* quark.

The electric charges of the quarks are very peculiar. The *u* quark has a charge of 2/3, and the *d* quark a charge of −1/3 (in units of the electric elementary charge determined by Robert Millikan).

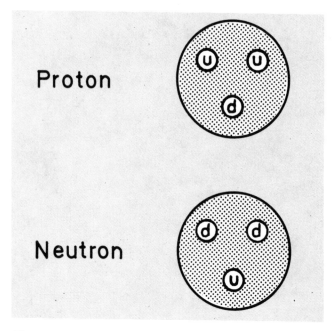

7.6 The structure of neutrons and protons.

$$\begin{pmatrix} u \\ d \end{pmatrix} \sim \begin{pmatrix} 2/3 \\ -1/3 \end{pmatrix}$$

Thus for the first time objects have been found in physics whose electric charges are not integers. The charges of protons and neutrons are nothing but the sums of the quark charges. For example: proton charge $= +1 = 2 \times u$ charge $+ d$ charge $= 2 \times (2/3) + (-1/3) = +1$.

Antiprotons and antineutrons are composed of the corresponding antiquarks denoted as $\bar{u}$ and $\bar{d}$:

Antiproton:	$(\bar{u}\,\bar{u}\,\bar{d})$
Antineutron:	$(\bar{d}\,\bar{d}\,\bar{u})$

What about π mesons? These particles, too, are composed of quarks, or, more precisely, of quarks *and* antiquarks. The positively charged π^+ meson is composed of a u quark and a $\bar{d}$ quark: $\pi^+ = (\bar{d}u)$ (see figure 7.7), and the negatively charged π meson is composed of a $\bar{u}$ quark and a d quark. Things become a bit more complicated in the case of neutral mesons. As can readily be seen, we have two possibilities for obtaining a neutral system out of a quark and an antiquark, namely $(\bar{u}u)$ and $(\bar{d}d)$. Which of these, if either, is the π^0 meson? The nonspecialist may find the answer given by physicists puzzling. The π^0-meson consists of both $\bar{u}u$ and $\bar{d}d$. There is a 50 percent probability for a π^0 meson to be a $(\bar{u}u)$ system, and a 50 percent probability to be a $(\bar{d}d)$ system. The neutral π meson thus is a hybrid.

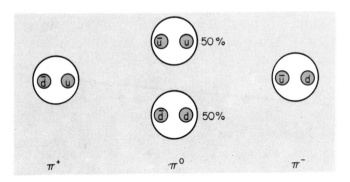

7.7 The internal structure of π mesons.

Mesons are composed of quarks and antiquarks, that is, of matter and antimatter. No wonder, then, that they are not stable, that they decay. This readily explains the decay of the π^0 meson into two photons: the quark-antiquark system annihilates into two photons, a process analogous to the electron and positron annihilation described earlier. Like it, the decay of the neutral π meson demonstrates Einstein's equivalence of mass and energy. The mass of the π meson is fully transformed into energy, that is, into the energy of the two γ quants emitted.

The life of the charged π mesons is also ended by annihilation, though not by the annihilation into γ-quants. The law of the conservation of electric charges prohibits this. Thus the π^+ meson, through the annihilation of u and $\bar{d}$, decays into a positron and neutrinos. The situation here is, however, more complicated, for, in an interim stage, still another particle, the electrically charged myon, is produced, which in turn also decays. But for the time being let us ignore these details. What matters to us here is that the charged mesons also decay due to the mutual annihilation of the quark and antiquark. A new interaction, called the weak interaction, which is also responsible for the neutron decay, plays an important role here.

In this context it should be pointed out that the quark model makes clear why π mesons are produced copiously in the proton-antiproton annihilation. For the quarks inside the proton it is simply a matter of uniting with the appropriate antiquark provided by the antiproton, and thus creating a meson.

A Dialogue about Quarks

This rough picture of our concept of the quark is sure to leave a host of questions unanswered. In order to simplify matters I will make you, the reader, an active participant in our discussion. You will pose questions, and I will try to answer them to the best of my ability. Here goes:

READER: One question popped into my mind when you first made reference to quarks, namely, why do three quarks unite to form a proton? Do the quarks exert some force on each other? Why this peculiar triple structure?

AUTHOR: You are quite right to ask these questions. When the quark model was first discussed some twenty years ago the very same questions arose. For almost ten years we did not know what impelled the quarks to form this *ménage à trois.* Today we know the answer. The quarks possess a unique characteristic, one not shared by any other particle, neither electrons nor neutrinos nor protons.

READER: And how does this special characteristic manifest itself?

AUTHOR: In physics, the characteristics of particles are often described by a charge. For example, an electron is surrounded by an electric field. That is what we mean when we say that the electron carries an electric charge. The neutron is not surrounded by an electric field; therefore it does not possess an electric charge.

READER: Even quarks have an electric charge. Does that mean that a quark is also surrounded by an electric field?

AUTHOR: Of course. As a matter of fact, electric forces of attraction and/or repulsion operate between quarks. A proton, for example, is composed of two u quarks and one d quark. The two u quarks carry an electric charge of 2/3. Since both these charges are positive, the quarks repel each other electrically.

READER: If that is so I don't understand why the quarks stay together at all. Wouldn't you expect the one *u* quark to be hurled out of the proton at high speed?

AUTHOR: That's precisely what would happen if the quarks possessed no properties other than their electric charge. That, however, is not the case. The quarks are not only electrically charged, but they carry still another charge, one that we refer to as *color*. Quarks can appear in three different colors: red, green, and blue.

READER: You must be joking. When I took chemistry in school, I learned that the color of a substance was related to its atomic structure. And now you are trying to tell me that color is an attribute of quarks.

AUTHOR: You are absolutely right. The color we are talking about here has nothing to do with actual color. Its only function is a descriptive one, to help make tangible still another property of quarks. Quarks, we have learned, are able to carry three different "charges" that have nothing to do with electric charge. These charges could, of course, be assigned abstract symbols, such as *a*, *b*, and *c*, or 1, 2, 3. In order to make things more vivid, physicists decided to use three different colors—red, green, and blue—to describe this phenomenon.

READER: You said that the colors, or, perhaps more accurately, the color charges, are independent of the electric charge. Does that mean that the electric charge is to be added to the color charges?

AUTHOR: Yes, a quark is characterized both by its color and its electric charge.

READER: What purpose do the colors serve? Are they related to the forces acting between quarks?

AUTHOR: Precisely. I already mentioned that the electric force between two electrically charged objects is generated because they are constantly "throwing" photons back and forth.

READER: I am beginning to think that quarks are constantly throwing something at one another.

AUTHOR: You're right. The objects that quarks are constantly throwing back and forth are called gluons. These gluons are so to speak the glue keeping the protons together (the term *gluon* is derived from "glue").

READER: Aha! The gluon is thus analogous to the photon.

AUTHOR: Not quite. The gluon, or rather, gluons, possess a characteristic not shared by the photon. When a photon reacts with an electrically charged particle such as an electron, for example, nothing much happens. Generally the only thing that changes in this process is the velocity of the electron, since the photon transfers its energy to the electron. When a quark reacts with a gluon, however, much more happens: the quark changes color. Thus its emission of a gluon can turn a red quark into a blue one. The gluon has carried off a red charge and brought in a blue one. Or to express it differently: it carries a positive red and a negative blue charge. If that gluon meets a blue quark it can transform that blue charge into a red one. In this process a red and a blue quark interact and exchange colors.

READER: What happens in the case of a green quark?

AUTHOR: A green quark can also send out a gluon, for example, a "green-be gone/red-come hither gluon." It is easily seen that there are altogether six types of gluons able to change the color of quarks. In addition, in three instances the colors do not change at all. We cannot, however, include one combination of these three in our count, for it has no effect: the red quark remains red; the blue, blue; and the green, green. Thus

7.8 The participants in the game of subnuclear forces: three colored quarks and eight gluons.

we have a total of eight different gluons, but only one photon (see figure 7.8).

READER: Hold on. I don't understand this business of the eight gluons.

AUTHOR: Every gluon changes a color into another one, although the two colors can be the same. We thus have these nine possibilities: red → green, red → blue, green → red, green → blue, blue → red, blue → green, red → red, green → green, blue → blue. Of the last three that leave the color unchanged, one is not taken into account, namely the one that leads each of the three colors back into itself. We are dealing then not with a total of nine gluons but with nine minus one, a total of eight gluons. The theory I am referring to is called quantum-chromodynamics (QCD). Developed in the early seventies by theoretical physicists, it is related to Maxwell's theory of electromagnetism: both theories are so-called gauge theories. (These are theories in which the interaction of particles is determined by rules formulated by the German mathematician Hermann Weyl more than sixty years ago.) The essential difference between Maxwell's theory and chromodynamics lies in the fact that in electrodynamics there is only one photon, whereas the forces in chromodynamics are transmitted by eight different gluons.

READER: That's all well and good in theory. But how do things look in practice? Have gluons and quarks been seen in laboratories?

AUTHOR: Until now there is no evidence that quarks and gluons do in fact exist as free, directly observable particles. And that's as it should be. For the theory of chromodynamics states that quarks and gluons cannot be isolated as free particles, that they exist only within protons and neutrons. Although quarks can be "seen" inside protons, for example with the help of the SLAC "microscope," they cannot be isolated from other quarks.

READER: Now I'm thoroughly confused. In principle it ought to be possible to extract a quark from the body of a proton. Why can't it be done?

AUTHOR: For comparison's sake, let's take a look at atomic physics. A hydrogen atom is a bound system composed of one proton and one electron. It is relatively easy to separate an electron from the atomic system, for example by irradiating the atom with electromagnetic waves of the appropriate length. This can be done because the force of attraction between the electron and proton in the hydrogen atom has a very specific characteristic; it becomes weaker the greater the distance between the electron and proton. As you may know, electric force diminishes in proportion to distance squared. A simple calculation shows that the electron in the hydrogen atom has to be given only a relatively small amount of energy (only about 14 eV) for it to be able to move away from "its" proton at will. Matters would be altogether different if the electric force would not diminish in proportion to distance squared. That's what happens in the case of quarks. Unlike the electric force, the force transmitted by the gluons (sometimes also referred to as chromoelectric force) does not diminish with the distance between the quarks; rather, at relatively large distances it remains practically constant. At least, we get this result when solving the equations of quantum-chromodynamics with the help of appropriate approximative methods; to date we have not succeeded in solving these equations precisely.

READER: Does that mean that you are not yet completely certain how the forces between quarks behave?

AUTHOR: Quite right; we are not altogether certain. However, there are many indications that the approximate calculations describe the situation fairly accurately. This calculation involves the use of big computers. The calculation effort is enormous, yet the result is quite simple: the forces between the quarks, at least in the case of relatively large distances—large in comparison to 10^{-13} centimeters—are constant. That means in particular that the force between the quarks at a distance of 10^{-12} centimeters is the same as at a distance of 1 centimeter—an astonishing result. In principle it is thus possible to put great distance between a quark and its two companions that originally were inside a proton— let us say as much as 1 centimeter (the normal distance between the quarks is about 10^{-13} centimeters). But this requires a large amount of energy, estimated to be about as great as the energy needed to lift a ton of material one meter. Anyone who has ever tried to hoist such a weight knows what that means.

Let me give another example. Let us assume we are going to move all the quarks in one liter of water so far apart that the average distance between them will measure 1 centimeter. To do so we would have to use an enormous amount of energy, comparable to the energy released in the explosion of hundreds of millions of hydrogen bombs.

READER: You just said that the distance between quarks in a proton amounts to only about 10^{-13} centimeters. How do you know this to be so?

AUTHOR: The typical distance between quarks is related to the size of the proton. Nuclear physicists have long known that the proton is not a pointlike particle but that it has a diameter of about 10^{-13} centimeters. Numerous experiments to determine the exact size of the diameter of a proton have been carried out, so that we can visualize a proton as being a small sphere with a diameter of about 10^{-13} centimeters. For a long time physicists were unable to find a satisfactory answer to the question of why the proton, unlike the electron, should have any size at all. Only with the emergence of the theory of quarks and of chromodynamics has light been shed on this question. We now know that the proton is composed of three quarks. Consequently it is not an elementary particle but an object of finite size whose diameter approximately describes the average distance of the quarks within.

READER: Still, I don't understand why the proton has a diameter of 10^{-13} centimeters rather than of 10^{-10} or 10^{-16} centimeters. What determines the size of a proton?

AUTHOR: Let me once more remind you of the hydrogen atom. The size of the hydrogen atom is determined by the electric force operating between the atomic nucleus and the electron as well as by quantum-mechanical uncertainty. Under normal circumstances the hydrogen atom has a diameter of 10^{-8} centimeters, neither more nor less. With the quark the situation is very similar. The uncertainty relations and the color force between the quarks determine the diameter of the proton: it is 10^{-13} centimeters. The diameter of every proton in the universe, whether here on earth or in a remote galaxy, measures 10^{-13} centimeters. Like the diameter of the hydrogen atom, that of the proton is a characteristic length prescribed by the forces of nature.

READER: There's a question to which I'd like an answer. It has to do with the "triplicity" of protons. Why are three quarks needed to build a proton? Wouldn't two be enough?

AUTHOR: No, that isn't possible, because the forces operating between quarks do not permit it. Because of their respective electric force, two quarks repel each other, just as two electrons do. For that reason two quarks cannot form a bound state, a particle. The color forces operating between the quarks show a remarkable feature. They cause the quarks to unite into color-neutral objects, to systems that show no color to the outside world. A proton does not consist of three quarks of whatever color, but of three quarks of three different colors—one red, one green, and one blue. This combination of colors can be called white, since the optical mixture of the colors red, green, and blue result in white. To the outside the proton thus shows no color; it is a white object (see figure 7.9). That is why the proton as an entity does not interact with the gluons, which react only with colored objects—with quarks or with other gluons. I would like to call your attention to an analogous situation in atomic physics. A hydrogen atom is composed of two electrically charged objects, a proton and an electron. The two charges balance each other. The hydrogen atom itself is electrically neutral.

7.9 Quarks come in three colors, and all three are found in protons. Each of the quarks in a proton is of a different color. The combination of the three colors results in a "white" object, a proton.

The situation for the proton is similar. It is composed of colored objects, the three quarks, but the proton itself has no color because the three colors of the quark cancel each other—the proton is "white." Now we can see why the proton is composed of exactly three quarks and not of two or four or whatever number. The simplest way to build a "white" object out of quarks is to combine three quarks of different colors; the triplicity of the structure of the proton is directly tied to the triplicity of the colors. Two quarks can never form a "white" system. Suppose we try it by taking a green and a blue quark. We are missing the third, the red quark, needed to construct a colorless system. The optical admixture of green and blue does not give us white. The same holds true for all other possible combinations. We simply must have three quarks involving all these colors to get a "white" object.

READER: But how about mesons? They are composed of two quarks.

AUTHOR: Hold on there. A meson is composed of a quark and an antiquark. That prefix "anti" is very important here. It is easy to produce a "white" object if one unites a quark and its corresponding antiquark, such as a red quark and the corresponding antiquark. In that case the colors balance each other and we get a "white" object, a meson. We thus have two simple methods for producing colorless structures. The one combines a quark and an antiquark of the same color, and the other, three quarks of different colors. Both these possibilities have been realized in nature, the first in the case of the π meson (in addition to the π meson there exist still other mesons, but these do not concern us here), the second in the case of the proton. We can now understand why there are no free quarks. Such a quark would not be a "white" object but a red or a green or a blue one. The rules of chromodynamics, however, do not allow this. It would take an infinite amount of energy to produce a free quark.

READER: How about the color forces between the quarks in a proton? If I understand you correctly we are dealing with forces far more powerful than electric forces.

AUTHOR: These forces are indeed more powerful than electric forces, but under certain conditions not much more so, at any rate not in the case of a quark in a proton. So long as the distance between the quarks is short, let us say about 10^{-14} centimeters, the forces are not particularly strong. They become so only if one attempts to separate the quarks. Physicists have found a name for this phenomenon. They call it *asymptotic freedom*. When the quarks are close to each other they behave like free particles. One might compare them to members of a chain gang. So long as the prisoners do not wander too far afield they can move about fairly easily. But if one of these prisoners were to get the idea of moving away from his comrades the bonds would tauten and impede his free movement. A proton is composed of three quarks bound by "chains," the color forces. The proton itself can move freely through space, but a quark inside the proton cannot. It must always follow its two partners. In chromodynamics the color forces between the quarks become stronger at greater distances (distances above 10^{-14} centimeters), until they reach a specific level; from that point onward the forces remain practically constant. This phenomenon to which we referred to earlier is called "infrared slavery." This designation is not politically motivated; in physics large distances are also called infrared distances sometimes.

READER: This infrared slavery can't be all that bad, since you said that the forces between the quarks are constant when the distances are great. Given sufficient energy, a quark could be moved from its partners far enough, at least in principle, for it to be studied as a single object. Admittedly, the amount of energy required makes such an experiment unfeasible, but in principle it ought to be possible.

7.10 When a quark and an antiquark are pulled apart a new quark-antiquark pair is created between the two particles. The result is two mesons that can easily move apart due to the absence of active color forces between them.

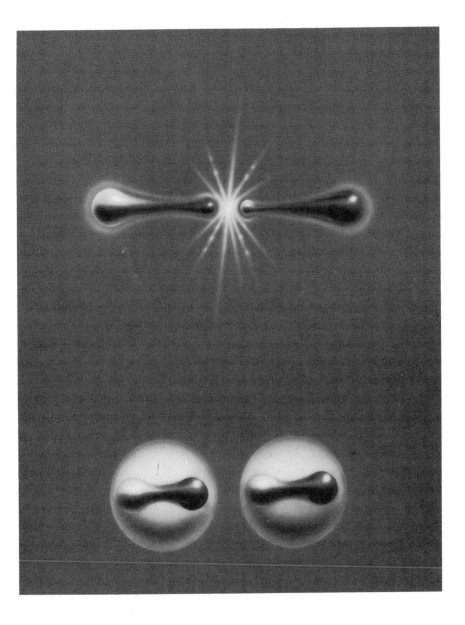

AUTHOR: Unfortunately I have to disappoint you. Even in principle this is not possible, and this is why. For the sake of simplicity let us take a meson as an example. Suppose we are strong enough to separate the quark and the antiquark in the meson, something that requires a great deal of energy. When we pull the two quarks apart (see figure 7.10) we pump energy into that system. We know that this energy can transform itself into mass, and vice versa. (A neutral meson, for example, can decay into two photons.) When we "pump" energy into our meson it is quite possible that we are producing new quarks and antiquarks, out of the void as it were. The energy expended by us changes into new quarks and antiquarks. Let us look at the genesis of a quark-antiquark pair. This new pair is created in the space separating the quarks. The chain that had united the two original quarks with one another suddenly breaks. The newly created antiquark combines with the old quark to form a meson, and the newly created quark unites with the old antiquark to form yet another meson. Our experiment has failed. Instead of putting more and more distance between the two quarks we produced two mesons. It is downright frustrating: all we ever get when we try to separate a quark from its partner or partners are new mesons. This thought experiment reminds me of the soap bubbles I used to blow as a child. You probably blew soap bubbles, no?

READER: Of course. But what do soap bubbles and quarks have in common?

AUTHOR: Let's say you have just blown a particularly large soap bubble. You then take a long knife and try to cut the slowly descending soap bubble in half. What happens?

READER: It can't be done. The bubble will burst. If you're lucky you'll get two smaller bubbles.

AUTHOR: That's exactly my point. A soap bubble can't be cut in half; the best you can hope for is two or more smaller bubbles. The same holds true for mesons. If you picture mesons as small soap bubbles you will understand that any attempt to cut them in half is bound to fail. All you can ever do is produce new mesons, new soap bubbles.

READER: On the surface this analogy seems convincing, but how do you know that quarks behave in this way? Has anyone ever tried to divide a meson in two?

AUTHOR: Such an experiment has been carried out with the help of the electron-positron annihilation. As you know, an electron and its antiparticle, the positron, annihilate each other.

READER: Yes, and as I recall, two photons are produced in the process.

AUTHOR: You are right in the case of electrons and positrons of relatively low energy. But when high-energy electrons and positrons collide frontally, other processes can also take place. For example, a quark and an antiquark can be produced out of the void, so to speak.

READER: Does that mean that quarks as free particles exist after all?

AUTHOR: No, it does not. When I say a quark and an antiquark are produced, I refer to the time immediately after the annihilation. This quark promptly combines with the antiquark, or with still another antiquark produced in this process, to form a meson.

READER: That's interesting. An electron and a positron accelerated to sufficiently high energies, when brought into collision, are able to produce a meson, or even a whole series of mesons.

AUTHOR: The process can be visualized somewhat like this: when the electron and positron collide, they destroy each other, producing a quark and an antiquark, either a $\bar{u}u$-pair or a $\bar{d}d$-pair. If the energy of the electron-positron pair was high enough, the two newly created quarks fly apart practically with the speed of light. If they are far enough apart, new quark-antiquark pairs are created out of energy, as shown in figure 7.10. This process continues, until at the end we see a series of mesons all flying in more or less the same direction, namely in the direction in which the originally created quark or antiquark took off. This configuration of particles is called a *jet.* Jets of this type have in fact been observed in electron-positron annihilations. The first time this was seen clearly was in 1979, in experiments carried out at the electron-positron storage ring PETRA at Hamburg. This apparatus is able to accelerate electrons and positrons to energies of about 20 GeV. (In 1980, a similar storage ring was completed in the United States. Known as PEP, it is located at the SLAC research center in California.) In the course of the electron-positron annihilation, two particle jets are often produced (see figure 7.11). There can be no doubt that these jets are produced in the process described above. The particle jets are akin to quark fragments produced during the electron-positron annihilation. Consequently, physicists call the production of particle jets the "fragmentation" of quarks. One might say that jets allow us to get an indirect view of quarks.

READER: If a quark has never been observed directly as a free particle, I wonder how one can justifiably argue for the existence of quarks. How is it possible for something to exist that cannot be observed in isolation?

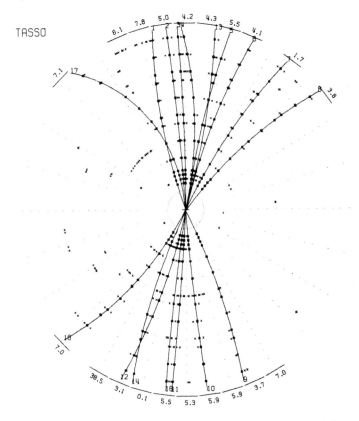

7.11 A typical two-jet occurrence in the electron-positron annihilation, as seen at the TASSO detector of the PETRA accelerator at Hamburg. The two particle jets are the "fragments" of the quark-antiquark pair produced in the electron-positron annihilation. (Photo: DESY)

AUTHOR: Here I part ways with you. What does "exist" or "observe" actually mean? No one has ever seen an electron directly either. Electrons can be observed only indirectly, by studying effects caused by them: for example, a track in a cloud chamber. Quarks have the unique property of existing only inside particles, not in isolation. But that is only a question of the scale with which we work. A physicist only 10^{-15} centimeters in height (unfortunately no such person exists) would be able to explore the interior of a proton. To such an observer quarks would be just as real as electrons are to us. Quarks are just as much objects of reality as electrons. Unlike electrons and protons, quarks cannot be isolated. To me that seems a new, interesting aspect that sheds light on the problem of our naïve concept of the divisibility of matter. Our daily experience teaches us that we can divide objects. We can split a log with an axe. Anyone who has ever chopped wood knows that it requires energy, however minute an amount relative to the energy hidden in the mass of the log being split, and for that reason it can be ignored. The energy required in the smashing of an atom or atomic nucleus is also rather small compared to the mass of the atom or nucleus. In the case of quarks, we are for the first time faced with a situation in which the energy needed for dissecting a proton or a meson into quarks is as great or greater than the mass of the particle. Our familiar concepts of the divisibility of matter are clearly in need of revision. Quarks are parts of a proton or a meson that can only be observed indirectly.

Having come to the end of our discussion I would like to sum up the major aspects of the physics of quarks. Protons and neutrons are composed of three quarks, namely of quarks of type u (charge 2/3) and type d (charge $-1/3$). Furthermore, there exist still other quarks, but they play no part in the structure of stable matter, since the particles composed of these quarks have a very brief life-span. For that reason I have not mentioned the interesting physics of the new quarks (for example, the *charm* quarks), of which you may have heard. Quarks carry "color." Because of the color forces operative between quarks, no isolated quarks can exist in nature. Quarks exist only within "white" objects (protons, π mesons, and so forth). The color forces are generated through the exchange of gluons. Gluons are colored objects like quarks, and therefore they too cannot appear as free particles. In closing I would like to mention nuclear forces, that is, the forces that cause protons and neutrons to form atomic nuclei. Today we know that these are not elementary forces like electric or color forces, but effective forces —indirect consequences of the color forces between the quarks. It is therefore not surprising that the nuclear forces that nuclear physics has been studying for decades are very complicated. Physicists began to interest themselves in particles because they wanted to understand the forces acting within atomic nuclei. The result is astonishing; they found a new world—the world of quarks, gluons, and color forces.

CHAPTER 8

Decaying Protons and the Unity of Physics

> Matter is composed of quarks and electrons. Quarks and electrons shall henceforth constitute a unit.

P R O T O N S and neutrons, the constituents of the atomic nucleus, being composed of quarks, are not elementary objects, whereas electrons, the building blocks of the atomic shell, are. At any rate, we have no evidence, either experimental or theoretical, to the contrary. The same holds true for neutrinos and quarks. Nothing points to a possible substructure of neutrinos or quarks. If electrons, neutrinos, and quarks were indeed made up of still smaller components, those would have to be squashed together in a very small volume. We have experimental proof that the diameter of electrons cannot possibly be much in excess of 10^{-17} centimeters, which means that electrons are either pointlike objects and have no internal structure, or that their diameter is less than one ten-thousandth of the diameter of the proton. Similar results have been obtained for quarks and neutrinos. The following explication and far-reaching speculations are based on the premise that electrons, neutrinos, and quarks possess no substructure, that they are elementary objects, although if it should turn out that they do, many of our conclusions may still be valid.

Are electrons, neutrinos, and u and d quarks related to each other? At first glance this question may seem odd. After all, electrons and u quarks differ considerably. For one thing, their electric charges are dissimilar; that of the electron is -1 (measured in Millikan's elementary charge units), while that of the u quark is $+2/3$. Furthermore, u quarks have color. They are either red, green, or blue. Electrons have no color. They are unaffected by the strong interaction. The differences between electrons and quarks are obviously great. Nonetheless, some years ago physicists advanced the hypothesis of a close link between electrons, neutrinos, and u and d quarks. We will not go into all the reasons for their hypothesis except for the one dealing with the electric charge.

Why is the electric charge of a u quark 2/3, or more precisely, 2/3 of the charge of electrons or positrons (ignoring their respective signs)? If there were no close links between electrons and quarks, we could not possibly understand this strange relationship between their respective electric charges.

Similar things can be said about the d quark. Why should its electric charge be 1/3 the charge of the electron? The electric charge of protons equals $+1$; it is composed of the sum of the charges of the three quarks. Our question can therefore also be formulated as follows: Why is the electric charge of a proton exactly the same as that of an electron? What would happen if their respective charges were not equal? That possibility defies the imagination. Atoms would then possess an electric charge and would repel or attract each other. Our world would simply not be the same. It clearly can be no accident that the electric charges of quarks are 2/3 and −1/3. What underlying reason is responsible for these charges?

How do we tackle the problem of establishing relationships between two different objects? We look for mutual properties. We have just pointed to one, namely the strange relationship between electric charges. Let us take this a step further and look at all elementary building blocks (including the color of quarks), and write these as follows (with r standing for red, g for green, and b for blue):

$$\begin{pmatrix} \nu_e & \vdots & u_r & u_g & u_b \\ e^- & \vdots & d_r & d_g & d_b \end{pmatrix}$$

We are dealing with a total of eight objects, six colored quarks, one electron, and one neutrino. We will henceforth use the term *leptons* to describe electrons, neutrinos, and their respective antiparticles. (The term is derived from the Greek *leptos,* meaning "light [in weight]." Compared to protons, electrons and neutrinos are extremely light particles.) The above eight leptons and quarks are called the lepton-quark family.

Suppose we take a closer look at this family. Is there anything particularly striking to attract our attention? Suppose we write down the respective charges of the leptons and quarks:

$$\begin{pmatrix} 0 & \vdots & 2/3 & 2/3 & 2/3 \\ -1 & \vdots & -1/3 & -1/3 & -1/3 \end{pmatrix}$$

What catches our eye is the fact that the sum of the electric charges of all the family members vanishes. Here is the simple calculation:

$$-1 + 3 \, (2/3) + 3 \, (-1/3) = 0$$

With this we have discovered another important relation between the electric charges of leptons and quarks. The sum of the electric charges of the members of the lepton-quark family vanishes. That cannot be sheer accident. There must exist a principle in nature to the effect that the sum of the charges must be zero. Let it be noted that the sum of the charges vanishes only if the three colors of the quarks are taken into account. Each quark appears in its three colors. If the colors were ignored, the sum of the lepton and quark charges would not be zero but rather $-1 + 2/3 + (-1/3) = -2/3$.

In order to understand why the sum of the charges is zero, physicists assume that leptons and quarks are related by a principle of symmetry wherein the leptons and quarks are nothing more than different manifestations of the same fundamental object, the same fundamental building block.

It is assumed, for example, that electrons and u quarks basically embody the same elementary object, that they are merely different manifestations of that object, the fundamental lepton-quark particle. This is not mere speculation. Rather, we are dealing with serious theories that can be tested experimentally. Two of these in particular have been closely studied by physicists. Without going into detail I shall try to outline their basic features. The first of these is the SU(5) theory proposed in 1974 by two Harvard University physicists, Howard Georgi and Sheldon Glashow. Other theories were discussed, in particular by Abdus Salam and Jogesh Pati.

According to the SU(5) theory, leptons and quarks can be derived from two fundamental particles. SU(5) is merely the mathematical term of the symmetry of leptons and quarks used in the theory.

Also in 1974, Peter Minkowski (now professor at the University of Bern), Georgi, and I discussed a theory involving only one fundamental lepton-quark particle. This became known as the SO(10) theory, because it involves a symmetry, which mathematicians describe by the symbol SO(10) (this is the symmetry of all rotations that can be made in a ten-dimensional space). What sets the SO(10) hypothesis apart is its assumption that all leptons, quarks, and their respective antiparticles are related, that they are simply different manifestations of one fundamental particle.

In the SO(10) theory, the lepton-quark particle can manifest itself in sixteen different ways, like a chameleon able to take on sixteen different colors. Here they are:

$$= \begin{pmatrix} \nu_e & \vdots & u_r & u_g & u_b & \vdots & \bar{u}_r & \bar{u}_g & \bar{u}_b & \vdots & \bar{\nu}_e \\ e^- & \vdots & d_r & d_g & d_b & \vdots & \bar{d}_r & \bar{d}_g & \bar{d}_b & \vdots & e^+ \end{pmatrix}$$

I would not like to give the impression that establishing a unified theory of leptons and quarks will solve all riddles of the universe. Many questions still remain unanswered. Nonetheless, we feel that theories like SU(5) and SO(10) accurately describe at least some aspects of nature. At this point I would like to mention the most important consequences of these theories.

Both theories, the SU(5) as well as the SO(10), state that leptons and quarks are related, but that this relationship normally goes unnoticed: it is frozen. The connection between leptons and quarks is noticed only when they are subjected to high energy whose order of magnitude can be estimated. It is enormous, somewhere in the 10^{15} GeV range, equivalent to 10^{15} proton masses (about 10^{-9} grams of matter). It is so vast that there is no hope of ever duplicating it in an accelerator. Only at an energy of 10^{15} GeV do all differences between leptons and quarks melt away. At that point there no longer exists a difference between electrons and u quarks, for example.

As a result of the uncertainty principle, the energy, or the momentum or velocity of a particle, is intertwined with the appropriate distance in space, which represents a limit for measuring the location of the particle. We can easily figure out the appropriate distance for the above-mentioned energy of 10^{15} GeV. It is about 10^{-29} centimeters, a minute distance. (Planck's elementary length is "only" a ten-thousandth of 10^{-29} cm.) At that distance all differences between leptons and quarks disappear. An electron seen at a distance of only 10^{-29} centimeters is no longer merely an electron but "only" a fundamental lepton-quark particle. At that point it is no longer possible to differentiate between electrons, neutrinos, and quarks.

How can that be, you may well ask, since the charges of leptons and quarks differ? And how about the color of quarks? Here, too, theory offers a straightforward answer: all differences between electric charge and color melt away. The electric forces as well as the color forces turn out to be "only" different manifestations of one and the same lepton-quark "fundamental force." The electric force we know from our daily lives is thus related to the subnuclear color force that induces the three quarks to form a proton.

It turns out that the relationship between electric force and color force comes into being only because in addition to these forces there exists still another force or interaction. Let us call it the X interaction, since physicists call the particles that transmit this new force X particles.

The X interaction possesses a most peculiar feature. For example, it is able to change a quark into an antiquark, and an electron into a positron. As though possessing magical powers, it is able to transform one member of the lepton-quark family into another. Here is how this transformation is accomplished. When we discussed color force we pointed out that the color of a quark may change when it emits or absorbs a gluon. Thus a red quark can turn into a green one. The X interaction follows a similar pattern. When a quark emits an X particle it automatically changes into another member of the lepton-quark family, perhaps into a positron. In this process the X particle transmits color and electric charge. For example:

$$\text{red } d \text{ quark } (-1/3) \rightarrow \text{neutrino } (0) + \text{red } X \text{ particle} \\ (-1/3)$$

(The parenthesized figures are the electric charges.) That means that the X particles, like the quarks, are colored objects and, like them, carry nonintegral charges.

Theoretical physicists introduced X forces in order to blend leptons and quarks into a unit. The SU(5) and SO(10) theories provide for only one lepton-quark fundamental force. The electric, color, and X forces are merely different manifestations of that fundamental force.

We still do not know whether X forces really exist. Experiments are under way. X forces are extremely weak, far weaker than electric forces. Theoretically they reach meaningful strength only when leptons or quarks are brought very close to each other, that is, to a distance of about 10^{-29} centimeters. Only at that small distance do the X forces become as strong as the electric or color forces.

The X forces behave somewhat like nuclear force. The latter is extremely weak and practically undetectable at a distance of more than 10^{-13} centimeters from the atomic nucleus. It becomes strong only when in the immediate vicinity of the nucleus. The respective distances, however, are very different. The critical distance of the nuclear force is 10^{-13} centimeters, and the critical distance of the X force, 10^{-29} centimeters. (For the benefit of those readers with some familiarity with nuclear force we would like to point out that the critical distance of the X force is related to the mass of the X particles. Electric and color force is transmitted by particles—photons and gluons— that have no rest mass. Not so X force. X particles do have mass, namely about 10^{15} GeV. In accordance with Einstein's equivalence of energy and mass, the mass of the X particles determines the critical energy at which leptons and quarks blend into a unit.)

In order for us to observe X forces we would have to bring two electrons, or one electron and one quark, very close together, to a distance of 10^{-29} centimeters. On the basis of the uncertainty principle, this would mean that both particles would have to be accelerated to extremely high energies, to about 10^{15} GeV. Obviously this cannot be done, at least not with the means available to us here on earth. Even if we were able to build an accelerator spanning the equator, we would still not be able, with the technological means presently at our disposal, to accelerate electrons to an energy of 10^{15} GeV.

Yet there exists another possibility to observe X forces, at least indirectly, and this brings us to one of the most interesting consequences of the unified theories of leptons and quarks: the instability of matter.

The Instability of Matter

Our world is full of unstable particles that are produced in collisions and decay shortly afterward, like the π mesons produced by the cosmic rays in the upper atmosphere that quickly decay and whose decay products continue to "bombard" the earth's surface (and our bodies). Neutrons, too, are not stable, but disintegrate within about 11 minutes into protons, electrons, and neutrinos.

Why does stable matter exist at all? Take a diamond. It is hard, seemingly indestructible, as though it would last forever. Formed here on earth billions of years ago, it has not changed perceptibly. It is composed of electrons and quarks living in peaceful coexistence now and apparently for all eternity.

Why has the diamond withstood the passage of time unchanged? The answer is to be found in the stability of the proton. If we observe a neutron long enough, it will decay into a proton, an electron, and a neutrino. But in a proton nothing happens. However long we observe it, it remains the same. It does not change into other particles. But why not? We know that during every elementary process in nature the electric charge cannot change. In the case of neutron decay, for example, the initial charge is zero (neutrons carry no electric charge), as is the charge of the end product, since the respective charges of the proton and electron cancel each other. If a proton were to decay, it would have to follow the law of conservation of the electric charge. This is quite possible. Let us look at all the decay possibilities of a proton. It could, for example, decay into a positron and a photon, or into a positron and a π^0 meson, which in turn decays into two photons (see

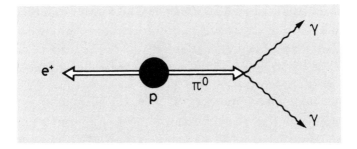

8.1 A hypothetical proton decay. The proton decays spontaneously into a positron and a neutral π meson, which in turn promptly decays into two photons.

---

figure 8.1). In these decays the electric charge of the proton would be carried off by the positron.

Scientists have been on the lookout for decaying protons, but none have been found. They apparently are stable particles. A proton could theoretically decay into a positron and a meson, but it simply refuses to do so. Experiments indicate that protons live for at least 10^{30} years.

The reader may ask how this figure was arrived at, since the earth is only about 5 billion (5×10^9) years old, and the experiments were conducted here on earth.

The thing to keep in mind is that in these experiments we do not observe just one, but many protons. We examine a macroscopic body of matter, for example a block of iron composed of about 10^{30} protons and 10^{30} neutrons. Even if the lifetime of a proton amounts to, say, 10^{30} years, an individual proton, in line with quantum physics, could possibly decay after an hour or a year, however small that possibility. Those are the sort of decays one keeps watching for.

Moreover, the existence of human life is proof that

protons enjoy a long life, in excess of 10^{16} years. Our body is composed of about 10^{28} protons. If protons live for 10^{16} years we might expect that about 10^{12} of these 10^{28} (10^{28} divided by 10^{16}) would decay in the course of a year, or about 30,000 per second. Since at every proton decay high-energy particles such as photons (γ quanta) are emitted, the human body would be subjected to a constant bombardment of particles that it could not resist for an extended period.

The unified theories of leptons and quarks make an interesting prediction: protons decay after about 10^{31} years. (The exact value depends on the specific model. In some versions of the SU(5) and SO(10) theories, the proton can live as long as 10^{32} years or slightly longer.)

We can visualize proton decay as follows. The above-mentioned X interaction is able to transform leptons into quarks, and vice versa, provided that the distance between the interacting particles does not exceed 10^{-29} centimeters. The probability for that to happen can be easily calculated with the help of the uncertainty principle. It occurs on the average of once in 10^{31} years. According to the laws of probability, this means that out of 10^{31} protons an average of one proton per year has the "chance" that two of its quarks will come as close as 10^{-29} centimeters. (If you want to visualize the amount of matter needed to supply you with 10^{31} protons, keep in mind that 17 metric tons of water contain about that amount of protons.)

When two of the proton's quarks come as close to each other as 10^{-29} centimeters, the new X interaction becomes effective. You will probably be able to figure out what then happens: one of the quarks will change into a lepton, perhaps into a positron. The life of the proton ends; it sends out a positron, or sometimes an antineutrino. What is then left is likely to be a π meson, and this, too, decays. It is thus predicted that protons will decay as described in figure 8.1. They are not immortal but have a lifetime of about 10^{31} years. The precise lifetime of the proton is not our major focus here. What is important is that protons do not live forever but eventually decay. Diamonds, too, are not forever.

The decay of a proton into a positron and two photons shown in figure 8.1 continues to interest us. In the process, a large part of the proton mass changes into photons, that is, light. Also, the positron is hurled out of the proton with practically the speed of light. This decay is an impressive example of the equivalence of mass and energy. The energies released during proton decay are enormous. If it were possible to induce all protons in 50 liters of water to decay in the course of a year, the energy released thereby would suffice to supply the United States with energy for a year. This, of course, is a utopian speculation. There probably exists no realistic possibility to accelerate proton decay sufficiently to meet our energy needs.

There may, however, exist a possibility of accelerating proton decay somewhat, a method reminiscent of the chemical mode of operation of a catalyst. The theories we mentioned—for example, the SU(5) theory—predict that there exist new, very heavy and very peculiar objects, the magnetic monopoles. These particles are very heavy, with a mass of about 10^{16} GeV, and they are able to transform leptons into quarks and quarks into leptons without the expenditure of energy in the 10^{15} GeV range. If such a monopole were to collide with a proton the latter would be done for. It would immediately decay into leptons and photons (as shown in figure 8.1).

A sufficiently large number of monopoles concentrated in a small volume would be the ideal instrument for the destruction of matter. For example, one might let a small amount of matter interact with monopoles. This would produce energy in the form of very intense radiation consisting primarily of γ quanta. A sufficiently large amount of matter interacting with monopoles would explode. Such a "bomb" would be more than two orders of magnitude more powerful than a hydrogen bomb.

The experimental quest for monopoles continues, but thus far without success. It is impossible to produce monopoles in accelerator experiments; the amount of energy available is insufficient. Only one possibility exists for discovering the monopoles; that involves a detailed examination of matter, particularly of cosmic rays. There is reason to believe that a large number of monopoles was produced by the Big Bang. If so, they must still be present in the universe, and it ought to be possible to find them with appropriate particle counters. Experiments are being conducted. Thus far none has been found.

I would like to emphasize that the density of monopoles in the universe cannot be very great. Under no circumstances can we expect to be able to use monopoles in the future for practical purposes.

What happens if protons indeed have a life of 10^{31} years? For all practical purposes it means they are stable. The earth would be losing only about 1/10 gram of matter per year through proton decay. The probability that a proton in the human body decays in the course of a normal human lifetime is about 10 percent; that is, only one out of ten of us will be lucky enough to "experience" a proton decay in the course of his or her lifetime.

For some years physicists have been testing the stability

of protons, in detailed experiments. In order to bar the distorting effects of cosmic radiation, these experiments are conducted underground, for example, in the Kolar gold mines in India or the Mont Blanc tunnel between France and Italy. In 1982 Italian physicists working in the tunnel discovered an event that may have been a proton decay (see figure 8.2).

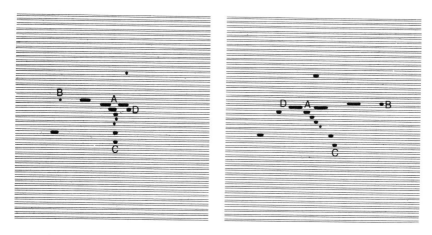

8.2 The phenomenon observed in 1982 by a particle detector constructed in the Mont Blanc tunnel. It is likely that it shows an example of proton decay, for instance the decay $p \to \pi^+ \pi^- \mu^+$ (the μ particle is a lepton with a mass of about 200 times the mass of an electron).

An unequivocal answer may not be found until results of the big new experiments now being carried out have come in. For example, a research team composed of scientists from the University of California at Irvine, the University of Michigan, and Brookhaven National Laboratory are engaged in testing vast quantities of water (10,000 metric tons containing about 10^{33} protons). The water is in a large basin 600 meters underground in the

8.3 The basin of the Morton salt mine before being filled with water. The basin is a large rectangle measuring $20 \times 27 \times 23$ meters. When filled with water it contains more than 10^{33} protons. Each of these protons has a chance, however small, to decay during the period of observation, which lasts a number of years. In decaying it emits a bluish light called the Cherenkov light, which is registered by one of the electronic devices affixed to the wall of the basin. (Reproduced by permission of the Brookhaven-Irvine-Michigan collaboration)

Morton salt mine east of Cleveland (see figure 8.3). When one of the protons of the water decays, rapidly moving particles such as photons or positrons (see figure 8.4) are emitted. These particles fly through the water and in the process produce a blue light. This phenomenon—the so-called Cherenkov effect—was first observed at the turn of the century by Marie Curie in Paris while she was working with radium. Radium is an element whose atomic nucleus is not stable; it decays (similar to neutron decay), and in doing so emits rapidly moving particles. The Soviet

8.4 Representation of a proton decay into a positron and two photons in the Morton salt mine. The three particles emitted during the decay produce three "light cones" of Cherenkov light, which are registered with the help of light-sensitive electronic cells.

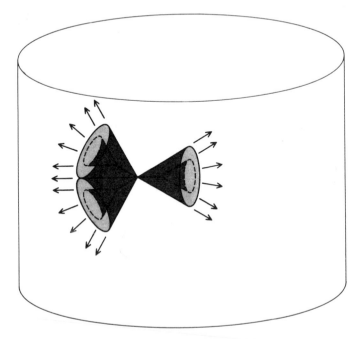

physicist Pavel Alekseyevich Cherenkov investigated the strange light phenomena seen in radium and other "radiant" elements in the 1930s.

In the Cleveland experiment the Cherenkov radiation produced during proton decay is registered by light-sensitive cells called photomultipliers. About 2400 such cells were installed around the basin in the Morton salt mine. Figure 8.4 shows what the decay of a proton into a positron and two photons might look like. Cherenkov's radiation is emitted in a characteristic conic shape. Since three particles are being emitted, three such "radiant cones" can be observed.

With the help of experiments under way in Cleveland and elsewhere, physicists should be able to see proton decay, provided that the life of a proton does not exceed 10^{33} years. The experiments started in 1983. To date no candidates for proton decay events have been found. The results of these experiments indicate that the simplest unified theory—the SU(5) theory—is probably not correct.

Cleveland is not the only hunting ground for decaying protons. Other experiments are being carried out or prepared in a number of locations, including Japan, the Silver King Mine in Utah, and a laboratory in the Bakzhan Valley in the Caucasus. Various teams of physicists in France, Italy, and the German Federal Republic are preparing a major experiment in the Mont Cenis tunnel under the Col de Frejus running from Grenoble to Turin. Over a period of years scientists using highly sophisticated electronics will monitor 1500 metric tons of iron.

The discovery of proton decay is bound to be of major significance, the irrefutable proof that protons, and hence atomic nuclei, do not live for all eternity, that they disappear in the course of time. Protons and atomic nuclei thus become objects with a past, comparable to the saurians that populated the earth for millions of years and then became extinct. What will happen to the world after the death of the protons in 10^{31} years? Does that mean the end of the cosmos?

Death and birth are closely linked. If protons ultimately die it is to be assumed that they were born in the remote past, in the infancy of the universe. The circle is closed. The insights of modern physics are closely linked with cosmology, with the origin of the world some twenty billion years ago. We will come back to the question of cosmology in the next chapters. But before doing so I would like to tell the reader about an astonishing thought experiment which is sure to prove very useful for the understanding of our cosmos.

CHAPTER 9

The Magic Furnace

Y O U may have seen the hot furnaces used in glass smelting in which glass is heated to extremely high temperatures until it melts. Now, in your mind's eye try to visualize such a furnace whose temperature can be raised to arbitrary high temperatures without itself melting down.

Suppose we had such a furnace with a 100-liter capacity at our disposal. Not fired, its temperature measures approximately 20°C. What does temperature mean? What happens when we heat an object? The atoms or molecules of an object generally are not at rest; they are constantly moving about. Thus every atom in the furnace wall oscillates around a specific point. The prevailing temperature determines the extent of these oscillations. As the temperature rises the oscillations increase in size, and vice versa. Temperature is simply a measure of the kinetic energy of atoms or molecules.

It is possible to cool objects so that their atoms cease to move altogether. When that occurs we speak of the absolute zero of temperature. Normally one uses the Celsius or Fahrenheit scale to indicate temperature. On the Celsius scale the zero point coincides with the temperature at which water freezes under normal conditions. This, however, is an arbitrary point, not related to absolute zero, which is equivalent to $-273°$ C. The atoms of an element are at rest at a temperature of $-273°$ C. Lower temperatures cannot be achieved.

Since the absolute zero of temperature obviously plays a special role it was decided to begin measuring temperature at that point. The Kelvin scale of temperature, named after the English physicist Lord Kelvin, does this. Zero degrees Kelvin hence means -273 °C (strictly speaking, absolute zero lies at -273.15°C, but for our purposes here we will round it off). Consequently, 293° K is equivalent to $+20$°C, the average room temperature. Under normal conditions water boils at 100°C, or 373°K. Henceforth, when we speak of temperature, we mean absolute temperature.

Let us assume that we succeeded in extracting all air from our furnace. It now contains nothing. Inside the furnace there is a vacuum. When I say "nothing," this is not strictly speaking true. The constantly moving atoms of the furnace wall continue to emit electromagnetic radiation, just as a hot radiator does. The radiation fills up the furnace's interior with photons.

One finds (in accordance with the physical laws of thermodynamics) that the properties of this radiation are determined by the temperature. The higher the furnace's temperature, the greater the average energy of the photons. Temperature and average energy are in proportion.

The number of photons also depends on the temperature. It increases by the third power of the temperature, measured from absolute zero. It thus follows that the total energy of the photon gas in the furnace, that is to say the sum of all photon energies, increases by the fourth power of the absolute temperature. When the temperature is raised by a factor of two, energy contained in the photon gas rises by a factor of $2^4 = 16$.

This law, incidentally, also applies to the radiation of an oven or radiator. A radiator in a room emits a stream of electromagnetic radiation, hence photons. The heat radiation of the radiator rises by the fourth power of the absolute temperature. It is easy to establish that a radiator at a temperature of 100°C emits 1.57 times as much energy in the form of heat radiation as one at a temperature of 60°C: $1.57 = (100 + 273)^4/(60 + 273)^4$.

What interests us primarily is the energy density of photon gas. It, too, follows the laws of quantum physics and thermodynamics. The figure we arrive at is easy to remember. At a temperature of 1°K, the energy density, or the energy per unit of volume, of photon gas in our furnace amounts to 4.72 eV/liter, or 0.00472 eV/cubic centimeter. To find the energy density at a higher temperature all we have to do is multiply the above figure with the fourth power of the temperature. For example: if the furnace is fired to a temperature of 1000°K, the energy per liter of the photon gas in the kiln will be $4.72 \times (1000)^4$ eV $= 4.72 \times 10^{12}$ eV $= 4720$ GeV. This energy is equivalent to the rest energy, or the mass, of almost 5000 protons.

Another magnitude of interest to us is the mean energy of a photon in photon gas. It is proportional to the temperature. For our purposes we can apply this simple rule: if we raise the temperature of the furnace by 1°C, the mean energy of a photon increases by 0.00008617 eV. If the temperature in the furnace measures only 1°K ($-$ 272°C), the mean energy of the photons equals 0.-00008617 eV. In this case we are dealing with electromagnetic radio waves. If the furnace has a temperature of 1000°K the mean energy of the photons will be 0.086 eV.

Now that we have laid the groundwork we can proceed with our thought experiment. In that context we will assume that the walls of our furnace are absolutely impenetrable and temperature-resistant, conditions which, of course, could not be found in reality.

We will now fire our furnace. For a long time nothing happens. The energy of the photon gas inside it continues to increase by the fourth power of the temperature. When the temperature climbs to about 6 billion degrees (6×10^9 degrees Kelvin) something strange occurs. As you will recall, the annihilation of a positron and an electron produces two photons. The same process can also take place in reverse; when two photons meet, an electron and a positron can come into being out of nothing, provided that the photons possess sufficient energy. The energy of each of the photons must be at least equal to the mass of the electron expressed in units of energy, that is, 0.511 MeV. As soon as the photons in our furnace reach a mean energy of a little more than 0.5 MeV, electron-positron pairs come into being. The temperature needed for this to occur can easily be figured out. It is about 6 billion degrees. At that temperature the photons in the furnace cease to predominate. They will exist side by side with electrons and positrons.

Incidentally, temperatures of that magnitude cannot normally be found in the universe. In the interior of stars temperatures reach "only" a few million degrees.

Suppose we raise the temperature in our imaginary furnace still more. As soon as it is sufficiently high compared to 6 billion degrees, the electrons, positrons, and photons begin to balance each other. There will be an equal number of electrons, positrons, and photons per unit of space. (This is only an approximation because of subtle differences between leptons—electrons and positrons—and photons.) The leptons and photons carry the same mean energy. Electrons and positrons are continuously being produced by the collision of two photons, and they continuously annihilate each other into two photons. On the average, however, there are as many photons as there are electrons and positrons.

Generally, but not always, the annihilation of an electron and positron gives rise to two photons. There also exists a subtle process called the weak interaction. (This interaction, which exists alongside electromagnetic and strong interactions in nature, is responsible for neutron decay.) It involves the annihilation of an electron and positron into a neutrino-antineutrino pair:

$$e^- + e^+ \rightarrow \nu_e + \bar{\nu}_e.$$

By this process neutrinos and antineutrinos are constantly forming in the furnace's bubbling "electron-positron-photon-stew." Let us assume that these neutrinos and antineutrinos are being deflected by the furnace walls, that they cannot escape. (Considering the highly unrealistic assumptions we have already made with regard to the nature of these walls, one more such assumption should not matter.) As a result we not only obtain a balance between the electrons, positrons, and photons, but the neutrinos turn into fully fledged equal partners. The furnace is now filled with electrons, positrons, neutrinos, antineutrinos, and photons.

Our heating process continues. The energy of the particles flying about in the furnace increases further, until it reaches a mean of 1 GeV, or a temperature of about 10^{13} degrees. At that temperature it is possible to create proton-antiproton or neutron-antineutron pairs. The collision of two photons, or of an electron and a positron, produces a proton-antiproton or a neutron-antineutron pair. Such processes are not mere theoretical speculations; they can be observed in accelerator laboratories like PETRA in Hamburg or PEP in Stanford.

When the temperature in our furnace begins to climb above 10^{13} degrees, the strongly interacting particles—the π mesons as well as the protons and neutrons—join in our game. From here on things become more complicated, for there exists a whole series of such particles. If I fail to go into this in greater detail it is because, as we shall find out, it would be pointless. At still higher temperatures the story again becomes quite simple.

Let us continue to raise the temperature in the furnace, this time to 10^{14} degrees. The particles in it now possess a mean energy of about 10 GeV. At this temperature something astonishing takes place. Protons and neutrons have disappeared from the furnace; instead, in addition to the photons, electrons, positrons, and neutrinos, we now find quarks and gluons. The mean energy of the particles has increased so greatly that protons and neutrons can no longer exist as independent entities. They disintegrate into their components—into quarks and gluons. These quarks and gluons are so dense that the mean distance between them is far less than 10^{-13} centimeters. In effect, the quarks and antiquarks and gluons inside the furnace behave like free particles, like electrons or neutrinos. All the particles are in a state of equilibrium. Electrons and positrons continue their reciprocal annihilation into photons or quark-antiquark pairs, while the latter annihilate themselves into gluons, into electron-positron or neutrino-antineutrino pairs. As a result of these continuous processes, the number of electrons, positrons, neutrinos, antineutrinos, photons, u quarks, d quarks, and gluons remains equal. (Subtle differences, such as those caused by the color characteristics of the quarks, do exist, but these amount at most to a factor of three in the various particle densities. Because for the time being I am concerned only with qualitative aspects, I am ignoring these factors here.)

And still we continue to raise the temperature in our furnace. Up to this point our imaginary experiment can be followed experimentally in our accelerator laboratories. The reactions relevant to my speculations, such as the annihilation of an electron and a positron into a quark and antiquark, are being closely studied in these labs. If I now raise the temperature in our furnace to more than 10^{14} degrees, I can no longer base myself on experiments, but only on theory. However, since in an imaginary experiment this is entirely permissible let us proceed.

For a long time, so the theory says, nothing much happens. At a temperature of, say, 10^{20} degrees, the particles in the furnace have a mean temperature of about ten million GeV. But the number of particles of all types remains equal. Finally we get up to a temperature of 10^{28} degrees, equivalent to a mean energy of 10^{15} GeV. As you undoubtedly recall, we came across this energy in the unified theories of leptons and quarks. It is determined by the mass of the hypothetical X particles, and it plays a role in joining leptons and quarks into a unit. As soon as we exceed the temperature of 10^{28} degrees the X particles also join in our game. At this point it becomes impossible to distinguish leptons from quarks.

Finally we find that our furnace contains a mixture of leptons, quarks, photons, gluons, and X particles. As a matter of fact, it is no longer possible to distinguish between leptons and quarks, nor between photons, gluons, and X particles. The former as well as the latter form a unit. At temperatures of more than 10^{28} degrees particles lose their individuality.

It might be useful to figure out the energy density in our furnace at a temperature of 10^{28} degrees. We find that it amounts to about 10^{103} GeV per liter, an energy density beyond the grasp even of physicists. The number of quarks and electrons in the observable universe is estimated to be about 10^{80}. If all these particles were compressed into a volume the size of a liter, the corresponding energy density would still be minuscule compared to the density in our initially empty furnace. Our experiment obviously is an imaginary one and must remain so forever. Still, that does not mean that we should stop. Therefore let us raise the temperature still higher, to about 10^{32} degrees. At that point the average energy of the particles will be somewhat greater than 10^{19} GeV. We now leave the realm of theoretical physics and enter unexplored, virgin terrain. At an energy level of 10^{19} GeV normal concepts of space and time collapse—at least according to quantum theory (see chapter 4). We do not know what takes their place, nor do we know what happens at temperatures of 10^{32} degrees or more. At these high temperatures or energy densities we probably can no longer speak of particles. The concept of particles, like the concepts of space and time, of energy and temperature, simply evanesces (see figure 9.1).

Temperature (in Degrees Kelvin)

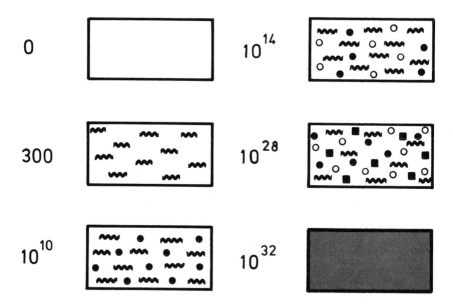

9.1 The magic furnace is heated. At 0°K the furnace is empty. As the temperature rises, for example to 300°Kelvin (27°Celsius), the furnace is filled with heat radiation—that is, photons. At a temperature of 10^{10} degrees the radiation is composed not only of photons but also of electrons and positrons (black dots). When the temperature reaches 10^{14} a plasma forms composed of photons, electrons, positrons, quarks, and gluons (the last-named are shown as open circles). At temperatures of 10^{32}, X particles join in (small squares). Nobody knows what happens at 10^{32} degrees. Our conventional ideas about space and time cease to be applicable.

The Magic Furnace

In our thought experiment we raised the temperature of the furnace from $0°K$ to 10^{32} degrees. When we began, the furnace was empty; by the time we finished, it was filled with highly intensive radiation composed of leptons, quarks, photons, gluons, and X particles (and still other particles which, for the sake of simplicity, we ignored).

We can also let this process take place in reverse. Suppose, beginning with a temperature of 10^{32} degrees, we slowly let the furnace cool off until it reaches a temperature of $0°K$, absolute zero. At temperatures below 10^{28} degrees, the X particles disappear from the furnace. What is left is a mixture of leptons, quarks, photons, gluons, and various other particles. When the temperature falls below 10^{13} degrees, quarks and gluons disappear, and all that is left are leptons and photons. Finally we reach the absolute zero of temperature. We look into the furnace and find that it is empty. During the process of cooling off, everything disappeared. Leptons and quarks annihilated each other; the surviving photons subsequently disappeared as we approached absolute zero.

Let us once more return to the hot furnace (which shows a temperature of more than 10^{28} degrees). At this temperature it contains leptons, quarks, photons, gluons, and, above all, X particles. We know that these X particles decay immediately upon birth, possibly into positrons and antiquarks: $X \rightarrow e^+ \bar{d}$. The anti-$X$ particles correspondingly decay into electrons and d quarks: $\overline{X} \rightarrow e^- d$. A continuous transformation process among leptons, q quarks, and X particles is taking place in the furnace. Let us take a closer look at these processes.

Strange asymmetries between particles and antiparticles can be found in nature. In some processes particles and antiparticles behave a little bit differently. One such effect was discovered by the physicists James W. Cronin and Val Logsdon Fitch, among others, in 1964, and earned Cronin and Fitch the 1980 Nobel Prize in Physics. Yet at the time, no one suspected that this discovery might hold a clue to our understanding of the universe.

Once again let us take a look at X particle decay. Some X particles can decay both into $e^+\bar{d}$ as well as uu; the corresponding antiparticles decay into e^-d and $\bar{u}\bar{u}$. One expects these decays to differ slightly from each other. For example, the $\bar{X} \to e^-d$ process should be more rapid than the $X \to e^+\bar{d}$ process. In the extreme, only the $\bar{X} \to e^-d$ and $X \to uu$ process might perhaps be possible. In that case an $X\bar{X}$ pair would decay into e^-uud, that is, into one electron and three quarks which at the appropriate low temperature could easily form a proton. With that, the decay of the $X\bar{X}$ pair leads directly to the production of hydrogen (see figure 9.2).

The radiation in our furnace is composed of a variety of particles and antiparticles. Thus there exists an equal number of X and $\bar{X}$ particles which can be grouped into $X\bar{X}$ pairs. Suppose we look at the furnace at a specific moment. We find that soon after, some of the X particles will have decayed. If every $X\bar{X}$ pair decays into e^-uud we can expect suddenly to find more quarks than antiquarks in the furnace. We have created quarks out of nothing, out of energy. Does that mean that we have created matter?

The answer to this unfortunately is no, we have not. Upon closer inspection it turns out that we have forgotten one important aspect. True, the X particles in the furnace continue to decay, but at the same time new ones are forming. We are faced with a situation which in physics is called "thermal equilibrium." Whenever we find such

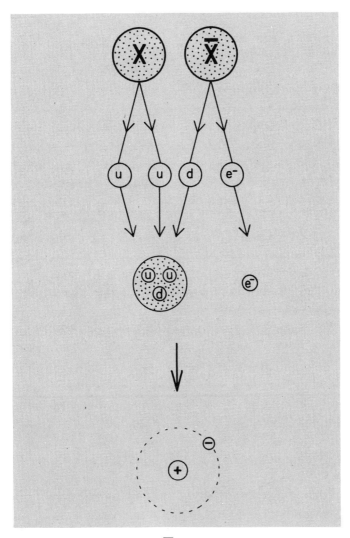

9.2 The transformation of an $\overline{XX}$ pair into three quarks (*uud*) and one electron. These are the components of hydrogen, the major constituent of matter in the universe.

a balance we also find complete symmetry between particles and antiparticles. More specifically, the number of quarks and antiquarks in the furnace remains constant. The only way we could create more quarks than antiquarks in the furnace is by destroying its thermal equilibrium.

This brings us to the last and most important stage of our thought experiment. We raise the temperature in our furnace to more than 10^{28} degrees and then try to cool it down as rapidly as possible. Since the furnace exists only in our imagination we are able to move it anywhere we want, to the remotest possible place, even interstellar space. And a good thing, too, because our next experiment is not altogether without danger. For what we are about to do is remove the walls of the furnace and leave the hot plasma with its temperature of about 10^{30} degrees standing free.

A furnace with an internal temperature of more than 10^{28} degrees must be able to weather a great many things. Its walls must not only be heat-resistant, but also be able to withstand enormous pressure. If we suddenly remove its walls the particles in the furnace no longer are confined. The hot plasma flies apart with the speed of light; an explosion takes place. An explosion is a physical phenomenon in which thermal equilibrium no longer exists. The particles squeezed together in the furnace now fly off in all directions. Each particle is on its own. Above all, the X and $\overline{X}$ particles decay after a short time, with practically no new ones being created, the thermal equilibrium having been destroyed. After these decays have taken place we count the quarks and antiquarks. The result does not come as a surprise: quarks outnumber antiquarks (see figure 9.3).

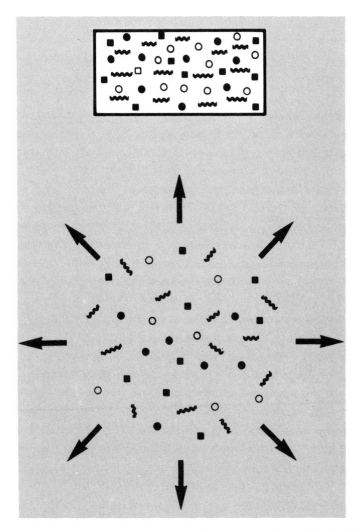

9.3 The furnace, heated to more than 10^{28} degrees, explodes. The "ash" of this explosion contains more quarks than antiquarks, and/or more protons than antiprotons. The matter in the universe is probably the product of such an explosion.

After the explosion of our furnace all unstable particles decay. Eventually the quarks and antiquarks unite and form either mesons, which also decay, or protons and neutrons and their respective antiparticles (the neutrons, however, in turn decay into protons and leptons). But at the end there will be more quarks than antiquarks, which means more protons than antiprotons. If we bother to gather up the residue left in the universe after the explosion, the antiprotons and their respective protons will annihilate themselves. Finally we will be left with some surviving protons that will unite with wandering electrons to form hydrogen atoms.

We have thus succeeded in producing matter out of "nothing," out of pure energy. It is the reverse process of the proton decay touched on earlier. In proton decay matter is transformed into radiation. Here we have done the reverse—we have made matter out of radiation.

Our mission is completed. We started with an empty furnace and heated it to temperatures of more than 10^{28} degrees. In the consequent explosion matter formed and we got hydrogen atoms.

The matter of the universe consists largely of hydrogen. What would seem more logical than to look at this hydrogen as the residue of an explosion that took place at temperatures of more than 10^{28} degrees? Are galaxies, stars, planets, and human beings composed of the ash of such an explosion?

CHAPTER 10

The Perceptible Universe

TODAY the earth lies spread out before our eyes in its entirety. It has become small. All we have to do is look at a globe to know exactly where we are located.

Thanks to the development of powerful telescopes the universe, too, has become accessible to us. Long years of astronomic research have yielded a globe of the universe. Facing it, we feel humility. Although comprehensible, the universe, measured by human standards, is vast. Our Milky Way contains 100 billion stars, and the cosmos encompasses at least that many galaxies.

Only three of these galaxies are visible to the naked eye. One is the Andromeda nebula, the other two are the Magellanic clouds, small galaxies visible only in southern regions. They are relatively close to us, only a little more than one hundred-thousand light-years away. The Milky Way, the Andromeda galaxy, the two Magellanic clouds, and a series of smaller galaxies called the local group are our neighbors in the galactic sea. The diameter of the local group is of the order of the distance between our galaxy and the Andromeda nebula, that is, about two million light-years.

With the help of powerful telescopes astronomers in the past fifty years succeeded in penetrating some billions of light-years into the cosmos. Thousands of galaxies have been closely studied. The number of galaxies in the visible cosmos has been estimated at more than a hundred billion.

As recently as a few hundreds of years ago only a select few could imagine the finite size of the earth. For most, the surface of our planet seemed infinite, beyond their imagination. Not so today. The earth has become small. Travel to remote countries has become commonplace. Television and radio have brought distant lands into our living rooms.

In recent years a similar change has taken place with regard to the cosmos as a whole. Today we are able to conceive of the cosmos as an entity. Its structure has become the subject of intensive research, and we are now able to plot a map of the universe.

When we travel long distances by car or on foot we think in terms of the number of kilometers or miles covered. But as regards the structure of the cosmos and the distribution of the galaxies, kilometers or miles are not very practical units of measurement. Neither are light-years, which are useful only for studying the structure of our galaxy, since the typical distance between two of its stars lies in the range of one light-year. But the average distance between galaxies is greater than just a few light-years. (For example, the distance between our galaxy and the Andromeda galaxy is about two million light-years.) I would therefore like to introduce a new unit of length, the MLY—a million light-years (1 MLY equaling 9.46×10^{18} kilometers). In the subsequent discussion of the structure of the universe I would ask you to think in terms of MLYs. Once you do, you will be surprised to see how simple the cosmos looks.

Measured in MLY the cosmos does not seem forbiddingly large. As we shall see, the observable part of the universe measures about 20,000 MLY. If you visualize the length of 1 MLY as 1 millimeter, then the observable universe measures 20 meters in diameter, the size of an auditorium measuring 20 meters $\times$ 20 meters $\times$ 20 meters. Let us call the identification 1 MLY-1 millimeter the cosmic length scale.

In units of MLY, our galaxy is barely visible. It resembles a 0.1 millimeter-large grain of sand. The distance between our galaxy and the Andromeda galaxy, 2 MLY, thus equals 2 millimeters.

Looking at the sky on a bright night we can see that the hundred billion stars of the Milky Way are not evenly distributed. Most of them are concentrated in the bright ribbon of the Milky Way. How about the galaxies? Are they, too, more highly concentrated in some regions of the sky than in others? Can we detect any notable structural patterns in the distribution of the galaxies?

Figure 10.1 shows the distribution of the galaxies in one area of the sky, the northern galactic hemisphere. This chart does not show all observed galaxies, but only those of a certain brightness—a total of about a million. The brighter a region shown here, the greater the number of galaxies in it. If we look at figure 10.1 more closely we notice the following:

1. The galaxies are not distributed absolutely homogeneously in the sky. There are bright and dark areas, somewhat like the uneven patterns formed by dye swirling in a glass of water. Some areas show a

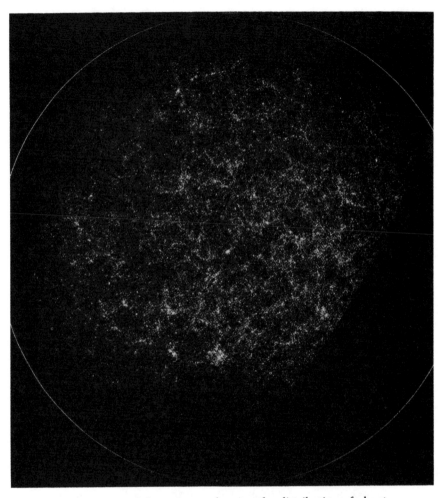

10.1 Overview of the universe showing the distribution of about a million galaxies in the Northern Galactic hemisphere, in one half of the sky. The lighter an area, the greater the number of galaxies found in it. The dense accumulation of galaxies in the center of the circle (the galactic equator) is the Coma cluster. Note the streaky distribution of the galaxies. (Prepared by P. J. E. Peebles, Princeton)

distinct galactic concentration, such as the Coma cluster in the center of figure 10.1 (see also figure 10.2). Our own local group of galaxies is also part of a bigger cluster of galaxies, the Virgo cluster, a concentration of some thousands of galaxies in the constellation Virgo, whose center is about 60 MLY (6 centimeters in the cosmic scale) from the earth (see figure 10.3).

2. Despite the observed fluctuations in the galactic density of the universe, no special structural distribution patterns have been found. This might lead us to conclude that on the average the galaxies in the universe are evenly distributed, like swarming insects. True, some regions contain a great number of galaxies, but then others house fewer. We also find vast cosmic regions that are empty, that contain no galaxies at all—holes with diameters of hundreds of MLY. In our cosmic length scale these holes are the size of footballs. Leaving such fluctuations aside, we can conclude that the galaxies "floating" in the universe are more or less evenly distributed.

Assuming this to be so, we can conclude that all galaxies, all spots in the universe, are of equal rank. There are no specially favored places. Every place in the universe is equal to every other place—"galactic democracy" reigns.

Let us put ourselves in the place of a remote astronomer, perhaps as much as 500 MLY away, who is looking at the cosmos through powerful telescopes. The picture of the galactic distribution seen by this astronomer will be very similar to that in figure 10.1. Only after detailed comparisons will differences emerge. If the observer were in the Coma cluster, for example, he would not see the

10.2 View of the center of the Coma cluster of galaxies. Only a few of the more than a thousand Coma-cluster galaxies are seen here. If the earth were located in one of the cluster's galaxies the night sky would be far more awesome than it is, and we would be able to see many of our neighboring galaxies with the naked eye. The distance between the earth and the Coma cluster is about 300 MLY (30 centimeters in our cosmic scale). (Photo: Mount Palomar Observatory, Pasadena, California)

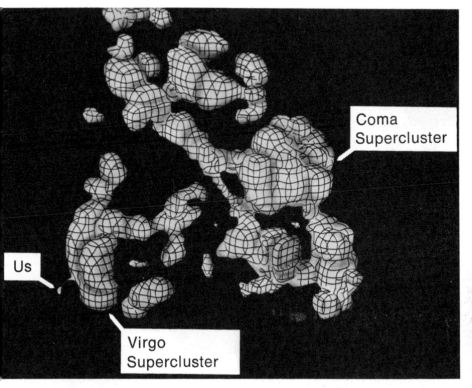

Coma
Supercluster

Us

Virgo
Supercluster

10.3 The distribution of matter in our cosmic neighborhood (galaxies, other types of matter such as neutrino clouds) might resemble this figure. Our local group is not far from a large concentration of galaxies called the Virgo cluster, which in turn is part of a system of many clusters of galaxies called the Virgo supercluster. Further away is the Coma supercluster of galaxies. (Prepared after a drawing by J. Silk, University of California, Berkeley, by WDR Television, Cologne. Reproduced with permission of WDR Television.)

strange concentration of galaxies in the center of figure 10.1, for he would be right in the middle of it.

We are now able to formulate a principle, one that astronomers call the "cosmological principle." It states that all observers in the universe are equal. The universe at large looks the same seen from any one point within it. Applied to our earth, the cosmological principle means that our own place in the universe situated inside the local group is not a special place at all.

The cosmological principle, modest though it may sound, has proved to be highly significant. Standing at the beginning of modern cosmology, a science that developed during the first half of our century, it was borne out by findings of astronomers at newly founded American observatories, particularly Mount Wilson at Pasadena and Mount Palomar near San Diego.

The Expansion of the Universe

One of the earliest scientists to venture into the exploration of space beyond our galaxy was the American astronomer Edwin Hubble. Soon after arriving in Pasadena in 1919, he and his devoted collaborator, Milton Humason, a former mule driver, turned to then still unexplored regions of the universe. Hubble also managed to get the most advanced observation and measuring instrumentation for the Mount Wilson Observatory. He was particularly interested in checking the measurements of the American astronomer Vesto Slipher, carried out around the turn of the century. What was at issue here?

Let us look at the light of a neon tube. As a result of

the perpetual transitions of electrons within the neon atoms between different states of energy, light—photons of a specific energy (on the order of 1 eV)—is emitted. The energies of these photons can be easily measured. Given appropriate instruments, physicists, by measuring the energy of the incoming photons, are able even at a great distance to differentiate between light produced by a neon tube and some other source, for example a candle. If that energy is equivalent to the normal energy of the photons emitted by neon atoms, the light source in question is beyond all doubt a neon tube.

Let us conduct yet another thought experiment. Suppose that you are generating a powerful jet of water with a garden hose, and that this jet hits your body with full force. You can feel its power. You then move toward the jet. The force of the water (its "energy") increases because the water hits your body at a higher speed (with greater kinetic energy) because of your own motion. You then turn around and move away from the water. The force of the water jet diminishes because the velocity of the water relative to you decreases.

The water-jet experiment can also be carried out with a light ray. It, too, exerts force when it hits your body, only the force involved is minute and cannot be felt. We therefore resort to another method, measuring the energy of the ray's photons with a spectrometer. To simplify matters we will use a light ray produced by a neon tube.

To begin with we examine the neon light when both we and the light source are at rest. We then move toward the light source and notice that the energy of the photons increases. The quicker our movement, the greater the energy of the photons. This increase in energy is manifested in a shift in the color of the light toward the blue end of the spectrum of visible light (the blue-light photons possess greater energy than the red-light photons). This effect corresponds to the increase in the force of the water jet when we move toward it.

If we move away from the light source the effect is reversed. The energy of the photons diminishes; they acquire a reddish hue. The same effect is achieved if we remain at rest and the light source moves either toward or away from us. In the first instance the energy of the photons increases; in the latter, it diminishes.

It thus becomes possible to measure the velocity of the light source simply by measuring the energy of the incoming photons. The speed of a galaxy or a star can similarly be determined by measuring the energy of the photons emitted by the galaxy.

Early in this century the American astronomer Vesto Slipher began to study the light of remote spiral nebulae. He discovered the effect that was to become known as the red shift. The light of remote spiral nebulae was shifted toward the red area. They appeared to be rushing away from us.

More than ten years later, Hubble embarked on the study of Slipher's effect, and in 1929 he finally made the discovery that pioneered the further development of cosmology. Hubble discovered that not only are the remote galaxies moving away from us, but that they do so more rapidly the further away they are. The observed velocities were by no means minor. Thus Hubble and Humason later found that galaxies in the constellation Ursa Major were moving from us at a rate of 42,000 kilometers/ second, that is to say, at one-seventh the speed of light.

Hubble proved that the velocities of recession of galaxies obey a simple law that can be described by a value known as Hubble's parameter H. The now generally accepted value of this parameter lies between 15 and 30 kilometers/second per MLY. Because many astronomers prefer the lower value, 15, it is the one we will use here.

According to Hubble's law, at a distance of one MLY a galaxy moves from us at a rate of 15 kilometers/second. At a distance of 2 MLY, the velocity doubles: 30 kilometers/second. In general, the velocity of recession is determined by the product of Hubble's constant, times distance in MLY.

A galaxy at a distance of 100 MLY would therefore have to move from us at a rate of $15 \times 100 = 1500$ kilometers/second. The above-mentioned galaxies registering a velocity away from us of 42,000 kilometers/second are thus $42,000 : 15 = 2800$ MLY, or 2.8 billion light-years away (2.8 meters in cosmic dimensions). Hubble's law is that simple.

It does, however, become more complicated when we try to interpret it. What made the galaxies decide to move away from us, and to increase the rate of flight the farther away they are? What about the cosmological principle of the galactic democracy? Is our galaxy perhaps a special case, and is that why the other galaxies flee from it?

You may find it reassuring to learn that Hubble's law does not apply to very small distances, that is to say to distances of only a few million light-years. The distance between us and the Andromeda nebula is a mere 2 million light-years. According to Hubble's law, it would have to be moving away from us at a rate of 30 kilometers/second. But in fact, the Andromeda galaxy and our own galaxy are moving toward each other at a rate of 300 kilometers/second. Presumably this is a consequence of the gravitational force between the two galaxies.

How is Hubble's law to be interpreted? Once again let us resort to an imaginary experiment by assuming that the earth will double in size within the period of an hour. All distances on its surface will double in size.

After the expansion time of one hour, the distance between New York and Chicago will not be 1000 but 2000 kilometers, and the distance between New York and Washington, D.C., will double from 300 to 600 kilometers. An observer in New York constantly measuring the distance between New York and other cities during the expansion time will find that all cities are moving away from New York at a rate proportional to the distance. Washington, D.C., for example, will be moving away from New York at a rate of 300 kilometers per hour, while Chicago will be moving much faster—at a rate of 1000 kilometers per hour. The observers in New York will find an expansion parameter (a "Hubble parameter") of 1 kilometer per hour per kilometer.

What impression do observers in Chicago get? They, too, find that everybody is moving away from them, and doing so at the rate described by the very same Hubble's law as that found in New York. For example, New York recedes from Chicago at a velocity of 1000 kilometers per hour. In short, every observer on the earth's surface encounters the same phenomenon—everything is moving away and the greater the distance of the place in question, the greater the velocity. Thus no observer is privileged above any other; all observation posts enjoy equal rights. A "geological" principle analogous to the cosmological principle is at work.

Let us relate what we just learned to the cosmos as a whole, to the flight of the galaxies, which can be interpreted as the moving apart of all galaxies. The galactic cosmos expands—the galaxies are flying apart. If that is so, they must at one time have been close together. When? Assuming that the velocity of recession remains constant or changes only very slightly in the course of time, the answer can easily be found. According to Hubble's law, a galaxy at a distance of 1 MLY is moving away at the rate of 15 kilometers/second. When did that galaxy depart from us? To find out we simply divide the distance (1 MLY = 9.46×10^{18} kilometers) by the velocity of recession of 15 kilometers/second to arrive at a time of 6.3×10^{17} seconds, or 20 billion years. Hubble's parameter thus gives us the time involved in the "expansion" of the galaxies. Assuming that the escape velocity has remained constant or changed only very slightly in the course of time, we can conclude that the galaxies were in close proximity about 20 billion years ago.

The time scale of 20 billion years is interesting, since it corresponds roughly to the time spans ascribed to the stars and the earth. The age of our planet is estimated at about 4 billion years, and that of the galaxy at approximately 15 billion. Does that mean that Hubble's parameter can tell us something about the birth of the universe?

Quasars

Perhaps quasars, the strangest objects yet found in the universe, hold the answer to this question. What are quasars?

We know with a fair degree of accuracy how much light is emitted by a typical galaxy such as ours. The greater the distance of a galaxy from the earth, the fainter its light.

That is why we cannot observe normal galaxies at a distance of more than about 1000 MLY. It is therefore easy to imagine the astonishment that greeted the discovery of objects more than 1000 MLY away and at times possessing a velocity of more than 250,000 kilometers/second: the quasars.

Quasars are very remote, relatively faintly glowing objects with a very strong red shift. But if the vast distance of quasars from the earth is taken into account, they are found to possess enormous luminosity. Some are more than a thousand times as bright as our own galaxy (see figure 10.4). That rules out the possibility that quasars are simply very luminous stars. But if not, what then are they? It is now generally accepted that quasars are very remote galaxies, or more precisely, the highly luminous nuclei of such galaxies.

The most convincing explanation for the enormous luminosity of quasars is that powerful explosions are presumably taking place in the nuclei, the centers, of these galaxies—processes by which gigantic amounts of matter are transformed into radiation energy. The simplest interpretation of this energy production is fantastic enough: that quasars are galaxies whose center is nothing but a big Black Hole that swallows up all surrounding matter—huge dust and gas clouds and entire stellar systems. In the process, enormous quantities of energy are emitted in the form of light and other electromagnetic radiation (such as X rays and radio waves). The reason that quasars are visible at all is due to this circumstance.

It can readily be seen that a quasar cannot afford this sort of energy expenditure for long periods, certainly not for billions of years. The time is sure to come when its supply of mass, and thus its energy, will be depleted.

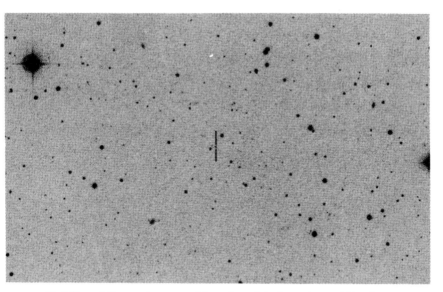

10.4 One of the most remote objects found in the universe. The object marked by a vertical line is not a star but a quasar (No. OQ172). It took light more than 15 billion years to travel from the quasar to the earth. At the time this object emitted its light the universe was only about one-fourth its present size. This quasar has long since been extinguished and transformed into a "civilized" galaxy, one possibly containing sentient beings. Perhaps our own galaxy also went through a quasar phase soon after its birth. (Photo: Mount Palomar Observatory, Pasadena, California)

Once the energy crisis catches up with it, its luminosity will drop to normal levels. This simple idea might explain why quasars are so far away from us, why none are to be found near the earth. A quasar "only" 200 MLY away from us would be visible to the naked eye. If the Andromeda galaxy were a quasar, we would be able to witness spectacular fireworks in the heavens.

Quasars probably represent an early stage in the evolution of galaxies, a galactic adolescence. The quasars we are seeing in our telescopes today no longer exist as the objects we are observing. Their light has been traveling for billions of years. When we see quasars we are looking into the dawn of the universe. We are seeing the cosmos as it looked billions of years ago. About 15 billion years ago, the quasar depicted in figure 10.4 was a highly luminous galaxy, but we have no idea what has become of it since. When it emitted the light we are seeing now the universe was about 5 billion years old and only about a fourth its present size.

Perhaps in its adolescence our galaxy also was a quasar whose light has been traveling through space for 15 billion years and today is seen in the telescopes of remote observers whose own galaxy appears to us here on earth as a luminous quasar.

The thesis that quasars represent galaxies in their adolescence is supported by the fact that almost 2000 quasars have thus far been studied, but none more than 18 billion years old has been found. It is thus safe to assume that in looking at the remote quasars we are glimpsing the threshold of the universe. If we were to look still more deeply into the cosmos, at the universe at a still earlier stage, we would not find any galaxies, for at that time galaxies in our sense did not yet exist. We are approaching the childhood of the cosmos, to which the next chapter is devoted.

CHAPTER 11

The Exploding Universe

> In the beginning God created the heaven and the earth. And the earth was without form, and void; and darkness was upon the face of the deep. And the Spirit of God moved upon the face of the waters. And God said, Let there be light: and there was light.
>
> —GENESIS 1:3

ALBERT EINSTEIN worked on his theory of gravitation—the theory of general relativity—from 1912 to 1916. It is beyond a doubt one of the most significant discoveries of the human mind. Space and time, so Newton told us, existed independently, not influenced by matter. They were absolute, God-given entities, so to speak.

Einstein's theory abrogated these Newtonian notions. According to the general theory of relativity, there is no such thing as absolute space and absolute time. Space and time are indivisible from matter. Their structure depends on matter. Space and time are a flexible continuum nestling against the available matter, be it a galaxy or a small planet. Space, time, and matter constitute a unit. Gravitation is a result of this mutual interaction of matter, space, and time.

This brief summary of the theory of relativity is all that is needed here; a deeper understanding of it is not absolutely required to understand the finer points of cosmology. However, we should keep in mind the major consequence of Einstein's theory, namely that space and time are not absolute quantities but are influenced by matter.

When he made his discovery and for some time thereafter, Einstein himself was not particularly interested in problems of cosmology, and so it came about that another scientist, the Russian mathematician Alexander Friedman of Leningrad, made the discovery bearing crucially on the development of cosmology. Contemplating the structure of the universe, Friedman asked himself whether it was infinite or possibly finite.

Once more we will take a look at the earth's surface and enlist it in one of our thought experiments (see also figure 11.1). The earth's surface is spherical in shape. It is a two-dimensional "space" (the two dimensions are characterized by the directions east-west and north-south); though finite in size, it is unbounded. If we start out anywhere on the earth's surface in whatever direction, we will never reach the end, but after a given period of time (ignoring oceans and other obstacles) we will find ourselves back where we started from. If we took a pedometer with us, we would know how many miles we covered as well as the circumference of the circle. If we divide this length by 2π we get the radius of the circle, and thus the radius of the earth.

The earth's surface represents a curved two-dimen-

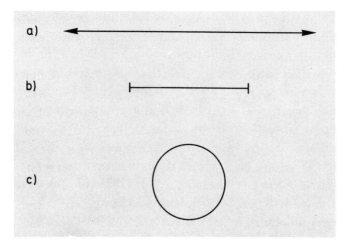

11.1 An infinitely long straight line (a) and a finite straight line (b) as examples of one-dimensional objects. (c) A circle is a one-dimensional space that is finite but unbound. A one-dimensional walker in that line will return to his starting point after a given time.

---

sional "space" that is finite yet unbounded. This is not a particularly profound discovery, since we know exactly what a two-dimensional surface of a sphere is. We are living in a three-dimensional space. Not only do we know the dimensions east-west and north-south, but yet a third one, up-down.

Suppose there existed two-dimensional creatures, some strange, amoebalike organisms without a third dimension, living out their lives on the two-dimensional east-west–north-south surface of the earth. And suppose these creatures decided to investigate the space in which they live. To their utter surprise they discover that their world is finite yet unbounded. In whatever direction they wander off, given enough time they inevitably will find themselves back where they started from.

Let us imagine an analogous situation in the three-dimensional sphere, a three-dimensional curved space. It would be a space without visible boundaries but finite in volume. Let us further imagine that we are living in such a space. How can we ascertain that this space is finite? By launching a rocket programmed to keep flying in one and the same direction. If the rocket returns to its starting point after a time, we are living in a finite space. The distance covered by the rocket can be taken to represent the circumference of a circle, and if we divide this result by 2π we will get the radius of curvature.

Applying Einstein's theory, Alexander Friedman investigated the structure of simple "curved" spaces (like the above-described analogue of a sphere). The result was surprising. He found that there is no such thing as a static universe, a space that does not change in time. In the course of time it either increases or decreases. The first case, the increase, would be in accord with the observed tendencies of galaxies to fly apart.

Friedman discussed two possibilities. In the first, the universe slowly expands until it reaches a certain volume and subsequently contracts again, like a balloon blown up to its maximum volume and then deflated. This would apply when the density of matter in the universe exceeds a given critical value, one that depends on the expansion velocity of the universe, that is, on Hubble's parameter. The universe thus is finite in volume but unbounded; it is a closed universe.

If the density of matter lies below that critical value, which will be discussed later, the universe expands endlessly. In this latter instance the spatial volume is infinite, like that of ordinary three-dimensional space. The distances in this universe increase continuously; that is to say, the galaxies fly apart forever. According to this theory there can be no return. We speak of an open universe.

The two possibilities discovered by Friedman can be made plausible by the example of the launching of a rocket ship. If we send up a rocket from the earth vertically we can distinguish between two different possible outcomes. The rocket can either achieve a high enough speed after burning off its fuel (more than about 11 kilometers/second) to escape the earth's gravitational pull and fly off into space or it can fail to reach the required speed. If that should happen, the rocket will ascend to a maximum height and then fall back to earth.

This latter possibility corresponds to the picture of a finite universe that first expands to its maximum volume and then contracts. Should the rocket disappear into space never to return, this would correspond to the possibility of an infinite universe that is forever "expanding"—a universe of galaxies forever moving apart.

Most of the properties of Friedman's two world models can be easily understood. In both, in the closed as well as the open model, the universe has a beginning, a time at which the expansion of space set in. It was, as we are about to see, a fiery birth. At that time the universe must have been extremely hot.

We can visualize this beginning only as a gigantic explosion—the Big Bang. The Big Bang, like any other explosion, hurled matter about. Its consequences are still apparent in the galaxies' ceaseless effort to pull apart. Friedman's theory predicts that eventually this "moving apart" will slow down.

Another look at the imaginary rocket might help explain the slowing down of the galactic escape. A rocket in a vertical ascent, having exhausted its fuel, will slow down, just like a stone hurled into the air. Both the rocket and the stone must "fight" against the pull of gravity, and in doing so, lose kinetic energy: they slow down. The matter hurled apart by the Big Bang also is subject to the force of gravitation. All galaxies in the universe attract each other, and consequently they must eventually slow down, because every galaxy must fight against the gravitational pull of its colleagues. In doing so each loses kinetic energy.

The reciprocal gravitational forces exerted by the galaxies depend on the number of galaxies per spatial unit. If the number is high, these forces might be great enough as to stop the galaxies in their tracks, even if only briefly, after which they continue to move toward each other, to "collapse" on each other. In that event the universe is finite in size (see figure 11.2).

If, however, matter in the universe should turn out to be sparsely distributed, the galaxies cannot be induced to turn back; they continue to rush apart, analogous to the rocket moving out of the earth's gravitational field. We are confronted with an open universe.

Today we are seeing a cosmos in which the galaxies are moving apart. Regardless of whether ours is an open or closed universe, we expect the galactic flight to slow down. Hubble's parameter describes the speed of recession of galaxies. That would indicate that Hubble's parameter is shrinking. At one time it must have been bigger than it is today. Hubble's parameter thus is not a constant number, but depends on the time interval since the Big Bang. That is why I call it Hubble's parameter rather than the less precise but widely used expression, Hubble's constant.

Despite intensive research it has not yet been possible

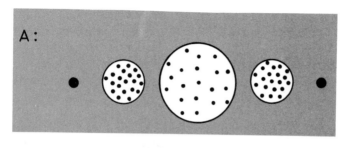

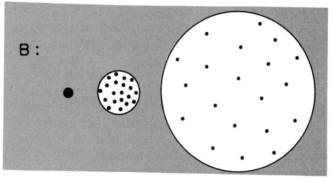

11.2 Two possible scenarios of the dynamics of the universe. (a) The universe expands, reaches a maximal size and subsequently collapses into itself. This occurs when the material density in the universe exceeds the critical density given by Hubble's parameter. (b) The universe continues to expand indefinitely. The material density is below the critical density. The universe is open, infinitely large. In fact, our cosmos is undergoing an expansionary phase. We are not certain whether ours is an open or closed universe.

to find a time dependence for Hubble's parameter. It decreases only very little, if at all, in the course of time (in accordance with a theoretical calculation based on the general theory of relativity).

What is the critical value of matter density in the universe, that is, the value that determines whether ours is a closed or open universe? It is determined by Hubble's parameter and the strength of the gravitational force. The latter is known exactly, whereas Hubble's parameter, as we have learned, is uncertain by a factor of two. If we use the 15 kilometers/second per MLY value of Hubble's parameter, we arrive at a critical density of 4.5×10^{-30} grams/cubic centimeter, which corresponds to about three hydrogen atoms per cubic meter.

What about the density of matter in the universe? We are living inside a galaxy. If we estimate the galactic matter density by counting stars, we obtain a mean matter density of about 10^{-23} grams/cubic centimeter (about 10 million hydrogen atoms per cubic meter). This figure obviously is far higher than the average density in the universe, for we have to take the vast empty spaces between the galaxies into account. When we do, we finally get a value of about 10^{-31} grams/cubic centimeter, which is only a small fraction of the critical value.

Does this justify the conclusion that the cosmos does not contain enough matter to meet the prerequisites for a closed universe? Is ours in fact an open universe? The answer is still uncertain. The cosmos contains vast amounts of matter, such as gas clouds, not visible in the form of stars or galaxies. We do not know whether the vast spaces separating the galaxies contain significant quantities of gas. It is furthermore possible that still other, hitherto hidden types of matter exist. We will return to this in the chapter dealing with the remote future of the universe. For the time being the question about the finiteness or infinity of the universe must remain unanswered.

Even if the universe should turn out to be infinite, the observable part of the cosmos is, of course, finite. The reason for this is connected with the finiteness of time since the Big Bang. If 20 billion years have elapsed since then, we on earth can receive only the light of remote objects that have been under way for at most 20 billion years. We cannot see a galaxy 40 billion light-years away because its light has not yet come down to us. If such a galaxy exists, it will not make its appearance on the horizon for another 20 billion years.

CHAPTER 12

The Echo of Creation

> It is even harder to realize that this pres-
> ent universe has evolved from an un-
> speakably unfamiliar early condition,
> and faces a future extinction of endless
> cold or intolerable heat.
> —STEVEN WEINBERG
> *The First Three Minutes*

A M E R I C A N industry is far more generous in its support of basic research than its European counterpart. Thus, Bell Telephone Laboratories has opened up opportunities for basic research in areas that often have had very little or nothing to do with its own product.

In the early 1960s two young radioastronomers at Bell, Arno Penzias of Columbia University and Robert Wilson of the California Institute of Technology, were involved in adapting a giant antenna at Holmdel, New Jersey, a structure shaped somewhat like an ear trumpet, for radio-communication with the first Telstar satellite. Penzias and Wilson also planned to "misuse" the antenna to explore the radiation of the Milky Way. Both these projects involved a close study of the antenna's sensitivity and the rather complicated receiver (see figure 12.1).

12.1 Robert Wilson (left) and Arno Penzias with their electronic "hear-ing aid," with which they discovered cosmic background radiation in 1965. (Photo: Bell Telephone Laboratories)

In the spring of 1964 the two scientists noticed a strange effect. At a wavelength of 7.35 centimeters they observed a relatively strong, peculiar radio murmur. Normally the radio signals they received with their antenna depended on its direction, but not so in this instance. These strange signals at a wavelength of 7.35 centimeters seemed to come from all over the sky, from all directions. The received radio signal (in the microwave region) resembled the radiation emitted by a body at a temperature of 3°K.

Where did it come from? Did this mean that the universe is filled with electromagnetic radiation, which is to say with photons?

One of Friedman's students in Leningrad was a man by the name of George Gamow, who in 1933 had left the Soviet Union for the United States for what he later said was "a more than thirty-year vacation from Soviet Russia." Gamow was particularly interested in the new fields of nuclear physics and astrophysics. It was he who first suggested that the universe at one time must have been much hotter than now—the notion of the "hot Big Bang." In 1948 (on April 1 of all days), a remarkable article about the Big Bang appeared in the journal *Physical Review* under the joint authorship of Ralph Alpher, a collaborator of Gamow; Hans Bethe, the eminent German-born nuclear physicist; and Gamow. Its sole authors, in fact, were Alpher and Gamow; Bethe had had nothing to do with the paper. But Gamow appended Bethe's name because he believed that a paper about so important a topic, the beginning of the world, ought to have as its authors persons whose names began with the first letters of the Greek alphabet: alpha, beta, and gamma. (It was the kind of inside joke from which none of Gamow's colleagues were safe.)

The $\alpha\beta\gamma$ theory was based on the premise that the universe in its primal state was a very hot mixture of nuclear particles, and that the Big Bang was the very rapid explosion of this hot plasma. The atoms we now find on earth and in the universe are simply, according to Gamow, the residue of that explosion.

The $\alpha\beta\gamma$ theory has long since become obsolete, except for one important result mentioned in passing by Gamow and his collaborators Alpher and Robert Herman in various papers (including Gamow's renowned work *The Creation of the Universe*). If the universe at one time had been extremely hot it would have contained a great many photons that would still be permeating the universe in the form of a homogeneous, isotropic radiation. This radiation, as estimated by Gamow and his collaborators, would correspond to the microwave radiation emitted by a body with a temperature of about 10°K.

We now know that Penzias and Wilson's strange radio signal is nothing more than the reverberation of the cosmic primal explosion predicted by Gamow, Alpher, and Herman. Subsequent detailed calculations by Robert Dicke and P. J. E. Peebles and others at Princeton (in part after Penzias and Wilson's discovery) support this finding.

At the time of their discovery, Penzias and Wilson did not take the astrophysical predictions seriously. They first were made aware of the importance of their discovery by an article by Walter Sullivan in the *New York Times* on May 21, 1965: "Scientists at the Bell Telephone Laboratories," it said, "have observed what a group at Princeton University believes may be remnants of an explosion that gave birth to the universe."

In 1978 Penzias and Wilson were awarded the Nobel Prize in Physics for their discovery. For Gamow (figure 12.2), who should have shared in the award, this recognition came too late. He had died on August 20, 1968, in Boulder, Colorado.

Penzias and Wilson's discovery, generally referred to as the Three-Degree-Kelvin Radiation, has since been tested in numerous experiments, including satellite experiments. It is a radiation emitted by a body with a temperature of 2.9°K, that is to say, a rather cool body. Such a body not

12.2 George Gamow shown constructing a model of the DNA molecule. (Photo: R. J. Gamow, University of Colorado, Boulder)

only emits radiation in wavelengths in the vicinity of 10 centimeters, as observed by Penzias and Wilson, but also electromagnetic radiation in wavelengths of 1 millimeter and less; the radiation is most powerful in the 1 millimeter wavelength. Unfortunately this radiation is almost completely absorbed by the earth's atmosphere. The observation of this part of the cosmic radiation involves costly balloon or satellite experiments.

Fortunately the human eye cannot see the cosmic background radiation, for if it could, the night sky, because of the residual Big Bang products that constantly stream down to the earth, would be bright rather than dark.

The energy of the cosmic radiation photons is minute. Thus, for example, the energy of the photons discovered by Penzias and Wilson, which corresponds to a wavelength of 7.35 centimeters, amounts to a mere 0.00002 eV. The photons of visible light, those registered by our eyes, have energies on the order of several electron volts.

We can imagine the universe as a vast container filled with photons of the cosmic background radiation (and, of course, with galaxies). Let us assume that this container expands. In that case the wavelength of electromagnetic radiation also "expands," and consequently the energy of the photons, and therefore the temperature of the radiation, decreases. The electromagnetic radiation we now find in the universe thus is nothing more than the "expanded" version of the hot radiation present shortly after the Big Bang. That would mean that the temperature of the radiation continued to fall as the universe evolved. In the past, when the cosmos was about one hundred-thousandth (10^{-5}) its present size, the photons of the cosmic background radiation possessed energies in the 1 eV range; they were visible to the eye (even though at the time, about 20 billion years ago, there was no one around to see them). Today the energy of the photons in the cosmos continues to decline, and with it the temperature of the photon sea. A billion years hence our descendants (if any) will be registering cosmic background radiation with temperatures of 2.75°K instead of today's temperature of 2.9°K.

Not only the energy of the photons but their density in space as well continued to decline as the universe expanded. Today's cosmic radiation with its temperature of 2.9°K possesses a photon density of almost exactly 500,-000 per liter, quite a substantial figure compared to the number of nuclear particles (the number of protons and neutrons) in the universe. As we learned in the previous chapter, a closed universe requires at least three nucleons per cubic meter, or 0.003 nucleons per liter. The matter visible to us as stars and galaxies, however, constitutes only about 2 percent of the critical density, or an average of 6×10^{-5} nucleons per liter. That gives us this relationship:

$$\frac{\text{Number of photons}}{\text{Number of nucleons}} \approx \frac{500000}{6 \times 10^{-5}} \approx 10^{10}$$

According to this equation the universe contains an average of 10^{10} (10 billion) photons per nucleon, a figure that is easy to remember and also, as we shall see, extremely important. Admittedly, the ratio given above has not been precisely determined, since the matter density in the universe is subject to numerous errors. We can, however, state with a fair degree of certainty that the number of photons per nucleon in the universe lies in the 1–10 billion range.

The photon/nucleon relationship is a characteristic number for our universe, a hallmark so to speak. Since in the expanding cosmos hardly any photons are produced or destroyed, and since the number of nucleons also remains constant, this relationship is constant in time, or at least has been since the time when the temperature dropped below the 1000 degree level in the infancy of the cosmos.

In chapter 9 we learned that matter can be produced by the rapid cooling of hot plasma. Our observations of the expansion of the universe and the photon sea lead us to believe that in its infancy the cosmos must have been extremely hot. It is therefore not farfetched to believe that the production of matter pretty much followed the pattern of the magic furnace. Let us now turn to the hot beginning of the cosmos.

CHAPTER 13

The Eightfold Way of Cosmic Evolution

> It is a development from the simple to the complicated, from unordered chaos to highly differentiated unity, from the unorganized to the organized.
> —Victor F. Weisskopf

HAVING LAID the groundwork, we are now in a position to reconstruct the evolution of the universe immediately after its birth. The first few seconds after the Big Bang must have been very lively, with lots of activity all around. Compared to these eventful times the universe today is a bore.

Following the method used in describing the cooling down of the furnace (chapter 9), we will divide the evolution of the universe into epochs. Measured by our human time scale, most of these epochs were extremely brief, mere fractions of seconds.

Our starting point is the primal explosion, which we will designate the zero point of our time scale, a time at which, according to the extrapolations of physicists, the prevailing temperatures were very high indeed, perhaps even infinite. In our discussion of the production of matter in chapter 9, I mentioned that we still are not quite certain how matter behaves at temperatures of more than 10^{33} degrees, since at such energy densities our normal concepts of space and time collapse. It is easy to estimate that in the time between the primal explosion and the ensuing 10^{-43} seconds, the temperature of the universe must have exceeded 10^{33} degrees. How should we describe the condition of the universe at that time? Was there even such a thing as time?

When one does not understand something it is best to keep one's mouth shut. In keeping with this principle there are those in the scientific community who suggest that it might be best to refrain from discussing the condition of the universe in its first 10^{-43} seconds and instead set the time of its beginning at the point at which the temperature dropped to below 10^{32} degrees. I am not one of these people; I believe that within the framework of a unified theory of matter and gravitation, we will eventually uncover the secret of the first 10^{-43} seconds.

What is needed to describe the evolution of the universe shortly after the Big Bang? If we accept the cosmological principle (see chapter 10)—and there is no reason to doubt its validity, at least for the first seconds or minutes of the universe—its evolution is calculable. Basically, the early universe was a rather simple object, whose condition essentially can be summed up by a single parameter: temperature. Once we know the temperature, we can figure out which types of particles play a role in the universe. As already indicated, the situation is similar to that of the magic furnace of chapter 9, except for one crucial difference: the volume of the furnace is fixed, while that of the universe is not. The universe is a system of finite or possibly infinite size that in its initial phase expands rapidly. The expansion velocity depends on the energy density, that is, on the prevailing temperature, and on the reciprocal gravitational force of its constituents.

Let us take a look at the rapid succession of epochs in the evolution of the universe.

FIRST EPOCH: THE MYSTERIOUS FIRST 10^{-43} SECONDS.

In the 10^{-43} seconds after the Big Bang the temperature of the universe exceeds 10^{32} degrees. Details about this first epoch are unknown.

SECOND EPOCH: 10^{-43} TO 10^{-33} SECONDS. MATTER IS PRODUCED FROM ENERGY.

This epoch begins after the first 10^{-43} seconds, at which time the temperature is about 10^{32} degrees. The universe is filled with a "primal stew" of all sorts of particles, including quarks, electrons, neutrinos, photons, gluons, X particles, and their corresponding antiparticles. The temperature falls precipitously. After 10^{-33} seconds it has dropped below 10^{28} degrees. The X particles decay and more quarks than antiquarks are left behind in the hot plasma. The excess quarks form the stuff out of which galaxies, stars, planets, and human beings will later form.

An important number should be mentioned here: the ratio between "excess" quarks (that is, the number of quarks minus the number of antiquarks) and the number of photons. It depends on the details of the X particle decay and can be estimated; typical values lie in the vicinity of 10^{-8} to 10^{-12}. To calculate this ratio we need precise information about the interaction of the X particles, which unfortunately we do not possess. That is why we are not yet able to establish the relation precisely but are forced to make do with crude estimates of the order of magnitude.

It should be emphasized that this quark "excess" is very minor. At the end of the second epoch the universe is composed of very hot plasma comprising quarks, antiquarks, photons, and other particles. At the time there existed about as many quarks and/or antiquarks as there were photons. For that reason the number of "excess" quarks compared with the total number of quarks or antiquarks respectively is very small, somewhere in the range of 10^{-9}. There is only one "excess" quark per one billion quarks and/or antiquarks. The "excess" of quarks left behind by the X particles plays no role for the time being.

THIRD EPOCH: 10^{-33} *TO* 10^{-6} *SECONDS. QUARKS COOL OFF.*

In this epoch the universe consists of hot plasma composed primarily of quarks, gluons, leptons, and photons cooling down from 10^{28} to less than 10^{14} degrees. Once the universe has completed this epoch, the particles of the "primary brew" possess an average energy of about 1 GeV (corresponding to a temperature of 10^{13} degrees).

FOURTH EPOCH: 10^{-6} *SECONDS TO* 10^{-3} *SECONDS. THE BIRTH OF PROTONS.*

As soon as the mean energy of quarks and gluons falls below 1 GeV, the mutual annihilation of quarks, antiquarks, and gluons sets in. Thus a quark and an antiquark can meet and annihilate each other and in the process produce two photons or an electron-positron pair. Likewise, two gluons can change into two photons. If the

226

universe were to contain the same number of quarks as antiquarks, the fate of both would be sealed. Neither would survive the annihilation after the first 10^{-6} seconds; all quarks and antiquarks would annihilate each other (as described in the slow cooling down of the furnace in chapter 9). Now the excess of quarks produced by the X particle decay becomes crucial. One out of about a billion quarks fails to find an appropriate antiquark with which to enter into a suicide pact. The remaining quarks have to accept the fact that they will have to bide their time in a progressively cooler universe.

We know that quarks are subjected to very powerful chromodynamic forces as soon as the distance between them grows sufficiently large. It is because of these chromodynamic forces that quarks get together in groups of three to form nuclear particles, either protons or neutrons, a process that is completed by the first millisecond (10^{-3} seconds).

What does the universe at that point look like? It is composed of a dense plasma of electrons, positrons, photons, and neutrinos, one dense enough to encompass about 10^{36} particles per cubic centimeter. In addition there are protons and neutrons—approximately 10^{27} per cubic centimeter. The average distance between the nuclear particles now amounts to 10^{-9} centimeters (corresponding to one-tenth of the extension of a hydrogen atom).

The dynamics of the universe are still determined by electrons, photons, and other particles that possess a mean energy of 30 MeV, and not by atomic nuclei, which contribute only a little bit, about one-millionth part, to the energy density of the universe.

FIFTH EPOCH: 10^{-3} TO 100 SECONDS. THE RADIATING UNIVERSE.

During this epoch the universe cools off to one billion degrees. A series of different processes follows.

1. Prior to this stage, the neutrinos were in constant interaction with all other matter. The universe was so tightly packed that even very weakly interacting neutrinos journeying through space at the speed of light were able to interact without any difficulty. After the first second, however, things change. The universe is still densely compressed, but no longer so dense as to guarantee regular collisions of neutrinos with electrons and other particles. Henceforth the neutrinos isolate themselves from the other particles and lead an independent existence—a state referred to as the decoupling of neutrinos.

2. At the beginning of the fifth epoch the universe contains the same number of protons as neutrons, since the latter had formed out of the u and d quarks. Because the number of u and d quarks initially was equal, the number of protons and neutrons also had to be equal. Neutrons, however, are somewhat more massive than protons (the mass difference amounts to about 1.3 MeV). As soon as the average energy of a particle (that is, the temperature in the universe) falls to a value comparable to the energy equivalence of the difference in the mass of neutrons and protons, a great many neutrons change into protons in processes involving electrons, positrons, and neutrinos (these reactions are similar to neutron decay). As a result, protons begin to outnumber neutrons. After the fifth epoch the universe contains about 75

percent protons and 25 percent neutrons. It should be stressed that neutron decay is not the agent responsible for the decline in the neutron population. At the end of the fifth epoch the universe is only 100 seconds old. The average life-span of a neutron is 11 minutes, which means that only a few neutrons decay during the first 100 seconds.

3. The most important process taking place during the fifth epoch is the annihilation of electrons and positrons, and it continues for some minutes after this epoch has finished. As soon as the mean energy of electrons and positrons falls below their mass (expressed in energy units), electrons and positrons annihilate each other, a process giving rise primarily to photons. Comparatively few electrons survive this mass extinction. They are the ones that ultimately will form the building blocks of the atomic shells. The total charge of the universe must equal zero, for only then does the certainty exist that the universe contains exactly one proton for each electron.

At the end of the fifth epoch the universe looks thus: it is composed of a photon and neutrino "brew" of about one billion degrees in temperature (detailed calculations show that the neutrinos are somewhat cooler than the photons, but this difference does not affect our qualitative observations). In addition, the universe contains protons and neutrons. Most of the electrons and positrons have annihilated each other in the course of the fifth epoch. However, every proton in the universe is matched by an electron, and consequently the total electric charge equals zero.

SIXTH EPOCH: FROM 100 SECONDS TO 30 MINUTES. NEUTRONS TURN INTO HELIUM.

After about three minutes the temperature of the universe falls to below 900 million degrees. In the meantime the number of neutrons continues to decline. The nucleons are composed of about 87 percent protons and 13 percent neutrons. Protons and neutrons collide quite often, remain together for a while, and form a bound system called deuterium. During this process excess energy is radiated off in the form of photons.

Deuterium is a relatively weakly bound system. Provided the necessary energy is at hand, it can easily be decomposed into its components—a proton and a neutron. If the temperature in the universe exceeds 1 billion degrees, deuterium has practically no chance of sustained survival. In its continual collisions with other particles, primarily photons, it is destroyed immediately upon birth. This situation changes only after the temperature falls to less than 900 million degrees, at which point the energy of the photons generally is no longer sufficient to destroy deuterium immediately after formation.

Deuterium particles collide continuously, and two of them can easily form a helium nucleus, an atomic nucleus composed of two neutrons and two protons. This atomic nucleus is unusually stable. It takes a considerable amount of energy to detach a proton or neutron from a helium nucleus (about nine times as much energy as is needed to split a deuterium particle). As a result practically all neutrons are captured soon after the onset of helium synthesis and consequently serve as components of helium nuclei. At the beginning of this process the matter formed by the nuclear particles is composed of 87

230

percent protons and 13 percent neutrons. Helium nuclei consist of two protons and two neutrons. It can easily be figured out that after the helium synthesis matter is composed of 77 percent hydrogen and 23 percent helium. This agrees remarkably well with the density of helium in the universe today. The cosmic background radiation as well as the formation of helium are thus direct witnesses of the early history of the universe, a time in which the cosmos was far hotter than now.

SEVENTH EPOCH: *30 MIiNUTES TO 1 BILLION YEARS. ATOMS FORM AND PHOTONS UNCOUPLE.*

After the first half-hour nothing much happens. The universe continues to cool off. Collisions between electrons, protons, helium nuclei, and photons are commonplace. But after about 300,000 years the universe enters unto a new phase—the age of the atom. The cosmos has cooled down so much that protons and helium nuclei are able to move about and capture electrons with the help of electric attraction. Atoms—hydrogen or helium atoms—form.

Atoms are electrically neutral. Since photons interact only with electrically charged objects, cosmic radiation and all the other matter in the cosmos begin to uncouple. At a far earlier stage the neutrinos in the universe have already made themselves independent; now the photons go through that same emancipatory process. After the passage of the first 300,000 years photons embark on an independent existence. The only thing that now happens to them is their progressive cooling off due to the expansion of the cosmos; they lose energy.

Today, after about 20 billion years, the photons have cooled down to a temperature of 3°K. The cosmic radiation observed by us is nothing but the "cooled" version of the intensive electromagnetic radiation that established its independence at the end of the first 300,000 years.

At the close of the seventh epoch matter in the universe is composed of hydrogen and helium atoms, but we also find intensive electromagnetic and neutrino radiation. From this point onward matter is on its own. Its composition and the forces operating between its particles will decide its future fate.

EIGHTH EPOCH: 1 MILLION TO 20 BILLION YEARS. MATTER CONDENSES. GALAXIES, STARS, PLANETS, AND LIFE DEVELOP.

After the uncoupling of the photons from the rest of the matter in the universe we enter the present phase of the cosmos: the matter-dominant phase. Matter, now in the form of hydrogen and helium atoms and only minimally in the form of photon and neutrino radiation, will from now on mark the expansion of the universe.

According to the Big Bang hypothesis, the major portion of cosmic matter, composed of quarks and electrons, takes the form of hydrogen and helium. This of course does not hold true for the earth, which is only a minor contributor to the matter of the universe. The earth is the result of a long evolutionary process. But when we examine the composition of matter in our galaxy, our findings correspond to the prediction of the Big Bang theory. About 22 percent of its matter is in the form of helium, and about 77 percent in the form of hydrogen. The contribution of the remaining elements is minor, on the order of 1 percent. About 0.8 percent of the mass of our galaxy is formed by oxygen, the most common element after helium. Iron, the most common metal, constitutes about 0.1 percent of the galaxy's mass.

The principal feature of the present, eighth epoch of cosmic evolution, is structure—a characteristic of the universe not found in the first million years. During its first seven eras the structure of the cosmos underwent little change. Energy and matter were distributed homogeneously; the universe was filled with uniform radiation. The only evidence of temporal evolution was found in the expansion of the cosmos and the constant drop in temperature.

At the beginning of the eighth epoch the universe was composed of a relatively hot gas made up of hydrogen and helium atoms. Today, shortly before the end of the second millennium of human time, about 20 billion years or 10^{18} seconds after the Big Bang, the cosmos is filled with galaxies, stars, planets, and such complex structures as we ourselves. The eighth epoch can rightly be called the epoch of structures (see figure 13.1).

About a billion years after the Big Bang, sizable regions in which matter is more strongly concentrated than elsewhere evolved out of originally minor variations in density. Through its gravitational force, this accumulation of matter influenced the matter surrounding it, causing still more matter to concentrate in the dense regions. Large clouds of matter composed of hydrogen and helium moved through the constantly expanding cosmos.

In the course of time these clouds concentrated—a consequence of the mass attraction operating within the clouds of matter. Most of these gas clouds began to rotate more rapidly about their axis, like pirouetting figure skaters with arms pressed to their sides.

Out of these hydrogen and helium clouds there evolved large, rotating accumulations of matter, the precursors of today's galaxies. Inside these gaseous clouds, matter con-

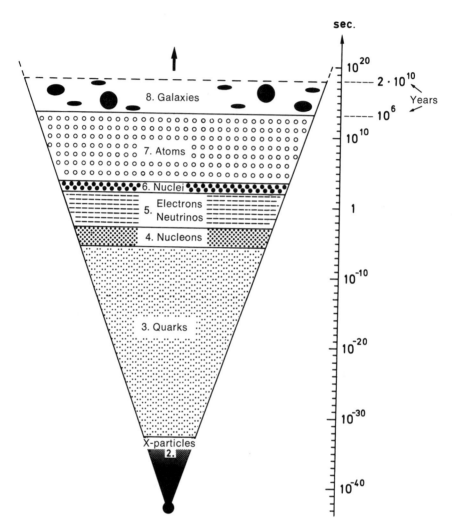

13.1 A schematic drawing of the evolution of the cosmos. Eight epochs of cosmic evolution are represented, beginning with the first 10^{-43} seconds following the Big Bang to the present epoch, which is marked by the presence of galaxies.

tinued to concentr as the small spherical structures formed. Th children itation condensed this
matter more andrger scale, ly the first thermonu-
clear reactions took place— usion of hydrogen nuclei
and atomic nuclei into other atomic nuclei: the first stars
began to illuminate the darkness of intergalactic space.

This evolution, which began about 15 billion years ago,
was to characterize the cosmos for many, many years to
come. It is marked by constant change, by the continuous
birth of new structures, as well as by constant decay, by
the death of surviving structures. It is interesting to note
that even structures as vast as galaxies go through a process of evolution. Shortly after their birth they contain
many massive stars that shine very brightly. That is why
younger galaxies radiate more powerfully than older
ones; some of them briefly live as quasars.

The first steps of the cosmos at the beginning of the
eighth epoch were quite powerful. The first stars did not
live very long. Most of them ended their relatively brief
existence in powerful explosions called supernovas. In the
process new matter for the formation of future stars—
nitrogen, oxygen, iron, the heavy elements—was created.
Much later the planets of the sun and we ourselves would
profit from the early supernova explosions. Much of the
matter on the earth, including our bodies, is composed of
the ash of these supernovas. Were it not for this residue
we would not be here.

About 4.6 billion years ago planets formed out of the
gas and dust of interstellar space. Less than a billion years
later, about 4 billion years ago, the first primitive organisms evolved in the oceans of that time out of complex
molecules that were continuously forming in these seas,
but which the highly energetic ultraviolet light of the sun

streaming down to the earth continued to annihilate. In a continuous evolution marked by the alternation of birth and death the first large cellular structures formed about 3 billion years ago.

One billion years ago, 3.6 billion years after the birth of the earth, the surface of our planet changed. Its atmosphere, composed primarily of oxygen and nitrogen, formed in interaction with the expanding organic matter, particularly the algae of the sea. Our atmosphere is largely the product of biologic evolution, a part of our biologic environment and, as such, vulnerable. Relatively minor local changes in the earth's surface, such as defoliation of the tropical rain forest, can have a devastating impact on all of the earth's atmosphere.

About 600 million years ago, in the Cambrian era, life exploded on earth. Many varieties suddenly sprang up: molluscs, conchs, sponges, starfish, fish, the first plants, insects. Two hundred million years ago dinosaurs arrived on earth. Their evolution came to a sudden and violent halt 60 million years ago, possibly triggered by a planetary catastrophe such as the collision of the earth with a smaller celestial body.

The extinction of the dinosaurs facilitated the evolution of mammals, which reached its peak in the evolution of human beings. This process, which began about 3.5 billion years ago, still has not run its course.

Dinosaurs dominated the earth for more than 100 million years, while mankind began its domination only some thousands of years ago. Will we survive as long as they did?

CHAPTER 14

The End of the World

> The world has been in existence for more than a billion years. As to the question of its end, my advice is: wait and see.
>
> —ALBERT EINSTEIN
> *Briefe*

IN the preceding chapter we found out that physics and astrophysics are able to describe the evolution of the universe since the Big Bang. Today the cosmos is in its eighth evolutionary stage. What next? All predictions about the future are uncertain and thus questionable, including predictions about the future of the universe. We do not have a clear answer; the best we can do is point to various possibilities, or, more accurately, to two alternate possibilities.

We know that the universe is expanding and that the remote galaxies are moving away from us. Hubble's parameter tells us the rate of this expansion, a phenomenon whose explanation lies in the Big Bang, the primal explosion. The remote galaxies are moving apart not because of a force that is pushing them away; rather they are moving apart as a result of the explosion about 20 billion years ago that we call the big bang.

Because of the gravitational attraction of the galaxies, their velocities of recession from each other decrease in the course of time. Is that slowing effect forceful enough to bring the galaxies to a stop and ultimately cause them to collide? Or will the universe continue to expand for all eternity? I shall attempt to answer these questions.

The deceleration of the velocity of remote galaxies depends largely on the mass present in the universe. The mass density of the universe required to put a stop to galactic flight, called the critical mass density, can be calculated. Its value, which depends on Hubble's parameter, is about 10^{-29} grams/cubic centimeter (equivalent to about ten hydrogen atoms per cubic meter).

This raises the question whether the existing mass density of the universe is bigger or smaller than the critical density. Only experiments and observation can give us the answer. We already mentioned that the mean mass density in our galaxy is about 10^{-23} grams/cubic centimeter. If we include intergalactic space, we arrive at a mean cosmic mass density of only 10^{-31} grams/cubic centimeter, a value below the critical value by approximately a factor of 100. However, the universe also contains dark gas clouds and other forms of matter that do not manifest themselves as luminous stellar matter. Their mass density, however, can also be estimated by various methods. This "dark" matter, it was found, might be contributing as much, and perhaps even more, to mass density as the luminous stellar material. Yet it is not possible to bring mass density close to critical density by the inclusion of this "dark" matter. The mass density of the cosmos would then equal only about one-tenth of the critical mass density, and possibly even less.

For the time being let us assume that the above assumption is correct and that cosmic mass density is in fact far smaller than critical density. If so, what does the future of the cosmos look like?

At first, nothing much would change. For some time, presumably even for many billions of years, galaxies and stars would continue to exist. Then gradually the galaxies would begin to dim. Nuclear reactions in the stars would slowly diminish, until finally the stars would be extinguished. The cosmos will grow cold and dark.

While all this is happening, nuclear matter proceeds to decay spontaneously into radiation. Finally, after about 10^{32} years, a considerable part of matter will have decayed. After 10^{40} years quarks will practically have ceased to be, which means all nuclear matter will have become depleted. Like the dinosaurs of early times, atoms have now become extinct.

What has not become extinct, however, are those strange objects mentioned earlier, the Black Holes. According to the general theory of relativity, Black Holes are static objects, singularities in the space-time fabric of the world, that remain unchanged in time: they are immortal. If that is in fact true, then Black Holes are the only cosmic structures that will survive.

But we have learned that Black Holes also do not live forever. They are not only black (not only in the sense that they are able to "swallow" any amount of matter, like a black body that absorbs light and changes it into heat); quantum theory tells us that Black Holes are "warm"; as they emit a constant stream of electromagnetic radiation, they lose part of their mass and grow progressively warmer. Finally they become so hot that they emit not only photons but also electron-positron pairs, and finally protons, neutrons, and their antiparticles.

Black Holes end their lives in an explosion. They evaporate and return to the universe the mass they had swallowed in the form of radiation (photons, electrons, positrons, and so forth).

The life of a Black Hole is very long, even measured by the time units we have been using. A Black Hole whose mass is equivalent to three solar masses has a life expectancy of about 10^{66} years. More massive ones live for correspondingly longer periods. Thus a Black Hole of a mass equivalent to the total mass of our galaxy is expected to live for about 10^{100} years. Even though Black Holes do not live forever, they obviously survive far longer than other matter in the cosmos.

The final, ninth epoch of the life story of the universe then will look like this: the constantly expanding universe is filled with radiation composed of photons and neutrinos that are continually losing energy. The cosmos is not likely to offer many interesting sights. It will be nothing more than a dark, inhospitable space, with only an occasional flare-up in the vastness of space—exploding Black Holes. In exploding they emit considerable quantities of electrons and positrons as well as protons and antiprotons. The latter fly off into space, but they also do not live forever. After about 10^{32} years the protons again change back into positrons, neutrinos, and photons. Now and then a positron traveling through space will meet up with an electron in order to annihilate into photons.

The cosmos does not now, nor do we think it ever will, contain Black Holes with a mass substantially greater than the mass of a galaxy. Thus after about 10^{100} years the ash of Black Hole explosions will also have disappeared. When that happens the cosmos will have come to a complete rest. Filled with barely perceptible, progressively cooler radiation of photons, neutrinos, electrons, and positrons, it sets out on its journey into eternity.

The above picture of evolution is based on the assumption that the observed mass density in the cosmos is smaller than the critical density. Thus far I have disregarded yet another possibility. Is it not perhaps possible that the universe contains far more matter than is found in the form of stars, galaxies, and so forth—matter that does not manifest itself in the form of protons and neutrons but in the form of other particles?

Another Source of Matter?

Let us once again take a look at neutrinos, those strange neutral particles that practically do not interact with normal matter. We have assumed that neutrinos, like photons, possess no mass, for the simple reason that up to now scientists have failed to find irrefutable evidence of such mass. In all past experiments neutrinos have behaved like particles with no mass. However, we cannot rule out the possibility that they contain some mass, however little. Mass values in the magnitude of 10 eV are being discussed, masses of about a fifty-thousandth the mass of an electron (10 eV corresponds to about 2 $\times$ 10^{-32} grams).

You might ask what role can so ridiculously small a mass play. Is it not a matter of indifference to the universe whether neutrinos have no mass whatever or a mass of 10 eV?

As we are about to find out, the answer to this is an emphatic no. Let us go back to the fifth epoch of the evolution of the universe. As you may remember, it began after the first microsecond and lasted for about 100 seconds. We mentioned that the neutrinos uncoupled themselves from the rest of matter. Since then there has been hardly any interaction between neutrinos and other particles in the universe, such as protons or electrons. After the first few seconds following the Big Bang, neutrinos, like photons later on, began to lead an independent existence.

If neutrinos had no mass whatsoever they would behave like photons. The universe would be homogeneously filled with neutrinos and photons. The cosmos contains on the average about 500 photons per cubic centimeter. According to the Big Bang theory, the number of neutrinos per unit of volume also lies in that range. More precise calculations indicate that the universe at present contains about 400 neutrinos and antineutrinos per cubic centimeter.

As we see, not only do we have a sea of photons, but a sea of neutrinos as well. The expanding cosmos causes neutrinos and photons alike to "cool off"; that is to say, in the course of time neutrinos continue to lose energy. It turns out that the mean energy of neutrinos in the universe is almost equal to that of photons (strictly speaking, neutrino energy is somewhat lower, but the difference is of no significance in this context).

If neutrinos possessed no mass, the neutrino sea, like the photon sea, would have practically no effect on the mass density of the universe. However, if neutrinos possess mass, however small, things look altogether different.

The crucial point is the fact that the universe contains a great many neutrinos, many more than protons, neutrons, and electrons. Thus if neutrinos possess mass they would add considerably to the mass density of the universe.

A quick calculation of the neutrinos' contribution to cosmic mass density yields this result: 6×10^{-31} grams/cubic centimeter $\times$ neutrino mass (in eV). Thus, given a neutrino mass of 10 eV, we would get 60×10^{-31} grams/cubic centimeter, or 6×10^{-31} grams/cubic centimeter, which is about equal to the critical mass density, and of course far greater than the density of normal nuclear matter.

If neutrinos have a mass of 10 eV, the neutrino sea's contribution to mass density is ten times that of the galaxies'—an astonishing result. Given a neutrino mass of more than 50 eV, the mass density in the universe would be greater than the critical density. In that event, the expansion of the cosmos would eventually come to a halt, braked by the gravitational effect of the neutrino sea.

With that the question of neutrino mass becomes central. Are neutrinos without any mass or do they possess some, and if so, how much? It is a question being pondered by physicists the world over.

In this connection I would like to call attention to the following:

1. According to the theory explaining electromagnetism, electromagnetic phenomena, and light in particular, photons, the particles of light, carry no mass. And experiments have borne this out. No such argument exists for neutrinos. Contemporary theories of leptons and quarks do not maintain that neutrinos cannot have mass. Some unified theories of leptons and quarks even suggest that neutrinos cannot be without mass, that they must carry

some mass (for example the previously alluded to SO(10) theory). Unfortunately the theories discussed are not specific enough to allow for clear-cut predictions for the neutrino mass. It should nonetheless be noted that crude estimates of neutrino-mass values range from 1 to 30 eV, figures which from a cosmologic perspective are highly interesting.

2. As previously mentioned, Soviet physicists in 1979 published the results of an experiment conducted in Moscow which allows the deduction of a neutrino mass of about 20 eV. However, there is reason to believe that this reading of the results is at least questionable.

3. Nuclear power plants emit a substantial portion of the energy achieved by nuclear reactions in the form of useless neutrino radiation. If neutrinos do possess mass, the neutrino emission of the nuclear reactors is expected to be slightly affected (one speaks of "neutrino oscillations"). In the early 1980s a research team headed by the German physicist Rudolf Mössbauer and the American physicist Felix Boehm examined the neutrino radiation of the Swiss nuclear reactor at Gösgen (see figure 14.1). Similar experiments were carried out by Fred Reines and his collaborators in the United States. No indications of neutrino mass were found, but on the other hand the possibility of a neutrino mass of 20 eV could not be ruled out either.* It may be some time before we will know beyond any doubt whether or not neutrinos possess a mass of 20 eV.

---

*It should be noted that in addition to the neutrinos studied in Moscow and Gösgen others also exist: the muon neutrinos and τ neutrinos. Since we do not know which (if any) contribute most to the mass density of the cosmos, reactor experiments cannot give us unequivocal answers to the question of their effect on mass density. Nonetheless, these experiments are highly interesting, for evidence of mass in the neutrinos emitted by the reactors would have far-reaching consequences.

14.1 The Swiss nuclear reactor at Gösgen whose neutrino radiation was tested for the effects of a neutrino mass. At the lower right, the shed housing the neutrino detector. (Photo: Swiss Institute for Nuclear Research, Villigen)

4. As we know, many galaxies do not live by themselves in the universe but form part of a cluster. A number of these huge galactic clusters, the Coma cluster among them, have been closely studied. In the process a problem known in astrophysics as the missing-matter problem cropped up. What does it signify?

In conformity with the law of gravitation, the galaxies within the cluster exert reciprocal gravitational force. They move relatively quickly to each other. If the gravitational attraction were suddenly turned off, the clustered galaxies would soon move apart, and the cluster would dissolve.

Gravitation, however, can keep a cluster together if that cluster contains enough matter. But this does not generally seem to be the case. In the early 1970s it was found that the galactic clusters then being studied all lacked sufficient matter. The galaxies in the clusters were not numerous enough to assure clustering. Some galactic clusters contain only a tenth of the matter needed for keeping them together. Ninety percent of the requisite matter is missing. Where is it?

If neutrinos possess mass, an elegant solution to this problem offers itself, one moreover which is being given serious consideration by many astrophysicists. I have mentioned the neutrino sea of the universe, analogous to the photon sea. What about this neutrino sea if neutrinos should turn out to possess mass? Would that mean that neutrinos are distributed as homogeneously in the universe as photons, that is, 400 per cubic centimeter?

In the expanding universe neutrinos continue to lose energy. As long as their energy in relation to mass is great, they move practically with the speed of light. But should mass overtake energy, this situation would change. If the neutrino mass amounts to about 10 eV, this change would have occurred some billions of years after the Big Bang, shortly before the birth of galaxies. Neutrino velocity would suddenly have diminished, and consequently, obeying the law of gravitation, neutrinos would have formed large neutrino clouds. Neutrinos would no longer be distributed homogeneously in the cosmos; instead, we would find some regions of relatively great neutrino density and others that are practically bare. The expanded, highly massive neutrino clouds must act like gigantic vacuum cleaners, sucking in the rest of cosmic matter—atoms and nuclear particles. According to this concept, a galactic

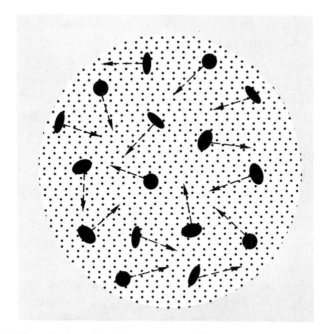

14.2 The galaxies of a cluster occasionally move at high velocities within the cluster (indicated by the arrows). Gravitation causes the clustering of the galaxies. Without it a cluster would disperse. Detailed studies show that the matter of the galaxies is not sufficient to hold a cluster together. If neutrinos possess mass, then a galactic cluster might be a "cloud" of massive neutrinos in which galaxies are "swimming about." The neutrinos would supply the missing matter (here indicated by dots).

---

cluster is simply a huge neutrino cloud in which the galaxies "swim around" like fish in an aquarium (see figure 14.2). The missing matter is provided by the neutrinos. Detailed calculations showed that it takes a neutrino mass of about 5–10 eV to achieve this effect. The neutrino density within the galactic clusters can be estimated at about 10^7 neutrinos per cubic centimeter, quite a substantial density.

At this juncture I would like to mention that our galaxy is part of a cluster. It is assumed that the area in which we live does contain about 10 million (10^7) neutrinos per cubic centimeter. Our earth, we ourselves, are thus swimming in a gigantic sea of neutrinos—a troubling, and for research physicists frustrating, thought. It challenges physical measuring techniques to come up with proof of the existence of this sea of neutrinos.

The possibility that neutrinos may possess a mass of about 10 eV leads to a curious conclusion. Neutrinos thus would constitute the major portion of matter in the universe. Normal matter—galaxies, stars, and planets—would contribute comparatively little to general mass density. They are, in a manner of speaking, the "waste" that is "contaminating" clean, invisible neutrino matter.

This ends our excursion into the realm of massive neutrinos. As can be seen, entirely new perspectives open up should neutrinos turn out to possess mass of about 10 eV or more. But if their mass should be less than 1 eV, they would not play a particularly significant role in astrophysics.

We cannot rule out the possibility that the mass density of neutrinos exceeds the critical density. Although this assumption is not supported by the astrophysical measurements of cosmic mass density, it is nonetheless possible.

Cosmic Contraction

If it should turn out to be so, what does the future of the universe look like? If mass density is greater than critical density, we are dealing with a finite, closed universe. For a while, the remote galaxies will continue to move away from us. Sometime in the future, perhaps 10^{12} years after the Big Bang, the cosmos will cease to expand. When that happens the galaxies will move toward each other, and the universe, in a reversal of its expansion, will begin to contract. All epochs of cosmic evolution will now pass by in reverse (see figure 14.3). Contemporary astronomers and physicists, if any, will find evidence for the contraction of the universe in the slowly rising temperature of the photon sea.

The energy of the cosmic photons finally will become so great that it will destroy stars and planets. Protons, neutrons, electrons, photons, and neutrinos once more

---

14.3 Two possibilities of cosmic evolution, described by the average distance between the galaxies in relation to time. (a) The expansion of the universe, which begins at zero time (Big Bang), ultimately ceases. Subsequently the galaxies collide and a reversal of the Big Bang takes place. In that event we are dealing with a finite cosmos. (b) After the Big Bang a never-ending expansion of the universe sets in; the cosmos is infinite in size. The admittedly imprecise measuring data regarding the distribution of matter in the universe favor the second projection. However, if neutrinos should turn out to possess mass the first scenario lies within the realm of the possible.

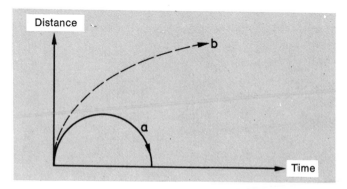

will form a hot gas, as they had shortly after the Big Bang. All macroscopic structures in the cosmos will be destroyed. Nothing will indicate that at one time the universe had contained planets inhabited by astrophysicists able to predict the future evolution of the cosmos. Finally the temperature of the universe will climb above 10^{28} degrees. The particles will begin to interact with the X particles produced in the collisions of leptons, quarks, antiquarks, photons, and other particles at the prevailing high temperatures. The X particles cause the last traces of the formerly cold universe, namely the small excess of quarks over antiquarks, to vanish. Just as this excess was once produced it now disappears. After the temperature climbs to more than 10^{32} degrees the universe enters the last epoch of its evolution, its final 10^{-43} seconds.

Like the first 10^{-43} seconds of the universe, the final 10^{-43} also are wrapped in darkness. We do not know whether the entire universe will contract to a point, to a singularity, as the mathematician calls it. If so, we can rightly say that time and matter will then cease to exist. The question What next? would become meaningless.

It can also happen differently. As a result of the still not understood effects of gravitation, the temperature may not continue its unending climb, but after reaching a maximum value may go down again because of a renewed expansion of the cosmos. A new cosmic cycle would set in. Once again, shortly before the temperature falls below 10^{28} degrees, excess quarks would be produced. This excess would act to form galaxies, stars, and planets in the eighth epoch of the new cycle. In this case nothing speaks against the possibility of an infinite number of such cycles. Our present epoch would then simply be the cold eighth epoch of one of the universe's perhaps infinite number of cycles.

The general theory of relativity tells us that the expansion of the cosmos, as determined by Hubble's parameter, will come to a halt if the cosmic mass density is larger than the critical mass density of about 10^{-29} grams/cubic centimeter. The observed mass density does not differ very much from this value; it is at most fifty times as small. This is a startling discovery, for essentially there is no reason why the mass density of the cosmos should in any way be related to the critical mass density. It could quite easily be a million times as small or as big. Some astrophysicists therefore believe that the cosmic mass density is in fact equal to the critical density. Such an equality may have been produced by an extremely rapid expansion ("inflation") of the cosmos immediately after the Big Bang (the so-called inflationary universe). If that is so, the finitude or infinity of the cosmos depends on subtleties in the distribution of matter.

This brings us to the end of our contemplation about the future of the cosmos. We will either enter a ninth, bleak, endlessly long, relentless epoch, a cosmic ice age, or the future universe will "revert" to the seventh epoch and then pass through all the preceding ones in reverse.

I have sought to give a picture of cosmic evolution. Even if its details should turn out to be inaccurate, there is still good reason to believe that evolution since the Big Bang did in fact follow the scheme outlined. That justifies our claim that we understand essential aspects of cosmic development, an understanding comparable to that about the evolution of the human species gained since Darwin's time. The discoveries of modern physics and astrophysics give us a rational understanding of the evolution of the cosmos since the Big Bang. The organisms living on earth today are not the only witnesses to a long process of evolution. The very stuff of matter is the product of a highly dynamic evolutionary process. Moreover, the very same processes responsible for the existence of matter will ultimately make for its disappearance.

Today we can say that we understand the evolution of the cosmos, but can we truly find an underlying meaning? Or is all cosmic evolution ultimately senseless? Is our existence merely a futile attempt by the cosmos to transcend its limits, to give itself a meaning that it does not and cannot have?

Science cannot answer these questions. Their answer involves value judgments, something science cannot offer. But by engaging in science and attempting to understand observed phenomena, we are also changing our way of thinking and acting. In this sense science is not free of values, which is why I wish to close this book with some thoughts on philosophy and religion.

CHAPTER 15

Unity and Diversity

> One feels as though melting into na-
> ture. One feels the insignificance of the
> individual even more strongly than
> ever and is happy about it.
> —ALBERT EINSTEIN
> *Briefe*

F O U R hundred years have passed since Galileo Gali-
lei, citizen of the Venetian Republic and professor at the
University of Padua, laid the groundwork for modern
natural science. In our century we have discovered the
structure of matter (see figure 15.1) and the forces of
nature that create this structure. Essentially three different forces hold the world together—the chromodynamic
forces operating between the quarks within nuclear particles; the less powerful electric forces within atoms; and
the all-encompassing force of gravitation. A remarkable
simplicity characterizes the world opening up to the scientist. Who would ever have thought that everything in
our motley, diversified world—galaxies and stars, rocks,
plants, and human beings—could be assembled out of
two kinds of quarks, u and d, and electrons? Nature, of
which we are a part, is highly complex; yet this complex-

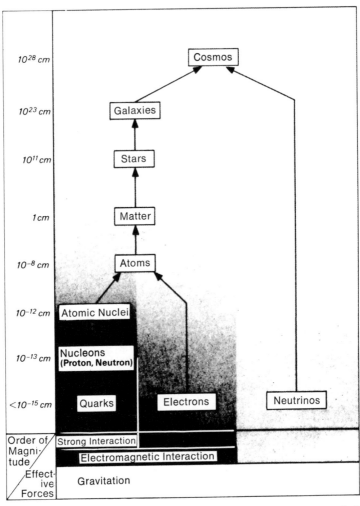

15.1 The structure of the cosmos is hierarchic. Matter is composed of quarks, electrons, and neutrinos. Strong interaction affects the quarks. Electromagnetic interaction is important both for the quarks and electrons. Elementary particles are affected by gravitation as well as by still another phenomenon, called the weak interaction.

ity is merely a variegated manifestation of one and the same basic matter.

A person weighing an average of 165 pounds is made up of these components:

u quarks: 7.0×10^{28}
d quarks: 6.5×10^{28}
electrons: 2.5×10^{28}

Simply combining these quarks and electrons will not, of course, yield a human being. We are more than merely the sum of our parts. The system *human being* is not a helter-skelter arrangement but a very specific combination of electrons and quarks that form hydrogen, oxygen, and carbon atoms, which, in turn, form specific molecules, obeying biochemical rules that developed during the earth's 4 billion years of evolution.

What makes the eighth epoch of cosmic evolution unique is the presence of structure. The first seven epochs after the Big Bang contained no notable macroscopic interacting structures. In that time the universe was filled with a hot gas composed of undifferentiated particles. An electron here on earth and an electron on the moon resemble each other even more closely than two eggs. An egg can at least be tagged. But electrons, quarks, and all the other particles cannot. The building blocks of matter are devoid of individuality. Boring uniformity is their fate.

The formation of larger accumulations of matter, the primal forms of the galaxies we know today, ushered in a special phase of the cosmos: macroscopic structures composed of numerous quarks and electrons began to form. (Some astrophysicists tend to believe that the seeds for this development were planted back in the seventh epoch by the formation of gigantic clouds of massive neutrinos. These neutrino clouds, by virtue of their gravitation, attracted the hydrogen and helium atoms that later formed, thereby giving rise to galaxies.)

What we are witnessing now is the interaction of those structures—be they galaxies, stars, planets, rocks, or living organisms—that began to form in the eighth epoch. For the first time the cosmos contained individual structures, identifiable objects. A rock is an individual structure. We can describe its shape, label it, and recover it at a later time. This may seem self-evident, but it is not. Individuality did not exist prior to the eighth epoch.

All structures that arose during the eighth epoch of the universe owe their existence to the fact that ours is now a cold universe. Only in a cold universe can macroscopic structures like galaxies and human beings exist. Moreover, these structures have not been created for all eternity. One day they will cease to be, and what will remain —increasingly warmer gas composed of particles or increasingly cooler gas of photons and neutrinos—will give no indication of the universe's former complexity.

How many individual objects, how many structures, can the universe contain? The number of quarks in the visible cosmos is estimated at about 10^{80}. The total number of possible objects—be they stars, planets, plants, or rocks—that can be made out of this building material, although finite, is so unimaginably large that for all practical purposes the possibilities may be said to be endless. The creative structural potential of nature is practically inexhaustible.

It is remarkable that nature, through the coherent cooperation of various constituents, managed to create new structures possessing astonishing, qualitatively novel characteristics. Many atoms, in appropriate combinations, are able to form a DNA molecule capable of reproduction. A musician may not find this surprising. With the few tones of the scale, the "constituents" of music, a wide range of melodies can be created, from pop songs to the most complex symphonic themes.

It is entirely possible that quarks and leptons are not the smallest building blocks of matter, that they are composed of still smaller constituents. However, today we can no longer deny that there exist very small, elementary constituents of matter. That means that the exploration of the detailed structure of matter is a finite process. One of these days in the not too remote future we will develop a closed theory of matter, a sort of "universal formula." But let me sound a warning note: such a theory will tell us absolutely nothing about the macroscopic, historical structure of our world. Many, perhaps even most, of the questions of interest to us cannot be answered by a universal theory, because such a theory is unable to take the highly important historical aspects of the eighth epoch into account.

This theory also fails to give us an overview of the existing diversity of objects and systems. For example, the fundamental equations of chromodynamics describe not only the behavior of quarks inside nuclear particles; they also describe atomic nuclei. Who would think that as a result of chromodynamic forces 168 quarks, or more precisely 82 u and 86 d quarks, would join to form a familiar object, the atomic nucleus of the element iron, which is composed of 16 protons and 30 neutrons. Nevertheless, the structure of the iron nucleus is embedded in the fundamental equations of chromodynamics. Thus the structure of the iron nucleus and, moreover, the structure of all atomic nuclei are "in principle" understood. However, it will never be possible to derive the details of the atomic nuclear structure directly from the chromodynamic equations. If we want to learn more about the structure of atomic nuclei we have no alternative but to explore their complex systems step by step in costly experiments and ceaseless theoretical work, the type of work nuclear physicists have been successfully engaged in for the past few decades.

Chapters 9 to 13 dealt with the origin of the universe and of quarks, atomic nuclei, atoms, and so on. We found out that these objects were structures that came into being during the cooling down of the universe. Even a lonely hydrogen atom wandering through space is a historical object. It carries within itself both the traces of its origin and the seeds of its destruction. This insight should help settle the ancient philosophic conflict between Becoming and Being.

Heraclitus summed up his philosophy by the dictum "Everything is in flux." He saw the world and all matter in a state of constant motion. Rejecting the notion of the immutability of structures, he thought them illusory.

Parmenides took the opposite view. For him, reality was a constant, and change nothing but illusion. Parmenides' philosophy gave rise to the notion that matter was composed of minute, indestructible units—the atoms of Democritus.

From our modern point of view both Heraclitus and Parmenides saw only part of the truth. Matter is composed of quarks and electrons, but the building blocks of matter will not exist for all eternity. They, too, are historical objects that came into being shortly after the Big Bang and that will disappear again after the eighth epoch. Creation and decay are as intrinsic to the constituents of matter as to all other objects in the cosmos.

Unity and Diversity

Nature in fact does not possess absolutely stable structures. To a chemist an atom may represent a stable system whose size no chemical reaction can alter. However, physicists can easily split an atom into its constituent parts. Nor are atomic nuclei stable. A π meson of sufficiently high energy colliding with an iron nucleus will cause that iron nucleus to burst apart.

The Unity of the Cosmos

There are no stable objects. Given a sufficient amount of energy every system can be broken up. I would therefore like to introduce the concept of "relative stability." Every object in the universe is a relatively stable system and therefore cannot last forever. This also means that such a system is in constant interaction with all other objects in the world. Nothing is or can be completely isolated. Even a proton moving through space millions of light-years from any galaxy "knows" that it is not alone. It is in constant contact with the world, a contact transmitted by the various physical forces and fields. This contact ultimately also is responsible for the proton's eventual decay. How would the proton "know" that it is not immortal were it not for its constant interaction with the rest of the universe? This interaction offers further proof that no object can exist in isolation from the rest of the world. The cosmos is not divisible—it constitutes a unity.

An important consequence of modern science is the renunciation of the naïve belief in the principle of cause and effect. Our brain is so designed that we are constantly falling victim to the illusion that every natural process is the result of the effect of a chain of linear causality, namely: this process takes place because of that previous process, and that process because . . . Thinking in terms of one-dimensional chains of causality is a human trait that describes reality incompletely. Every object in the cosmos has a multitude of relationships with its surroundings. One or more causal chains cannot adequately describe that object's dynamic behavior.

When I speak of the unity of the cosmos I also have in mind that ramified network of relationships encompassing every system of the universe. This network is a typical product of the eighth epoch made possible by the diversity of that epoch.

No such network existed prior to that time. An account of the evolution of the cosmos during the first seven epochs is a simple affair. It can be done by using thermodynamic notions like temperature. The evolution of the cosmos from the Big Bang to the end of the seventh epoch can be characterized by a one-dimensional chain of causality. But any fairly comprehensive account of the ramified network of relationships between objects in the eighth epoch must rely on the correspondingly complex methods of such sciences as cybernetics and systems analysis, which are especially suited for describing that complex network of interrelationships.

The Role of Scientists

The example of quantum theory has shown us that our conventional concepts are rather crude instruments for describing the subtle processes of atomic physics. Moreover, we cannot make exact predictions about individual processes. All we can do is predict probabilities. That is why quantum physics describes an open universe—nothing can be predicted with absolute certainty. This also means that the findings of the natural sciences are to be considered approximations. An idea becomes a scientific idea only if it can also turn out to be wrong. Wolfgang Pauli once commented after a lecture by a fellow scientist: "Your hypothesis is so faulty that it cannot even be wrong." A devastating critique.

Scientific research today more than ever implies not only the investigation of existing natural phenomena, but also the creation of new phenomena. We are building computers and sophisticated accelerators, we produce new chemical materials and a great deal more. At the same time scientists put their stamp on what they are investigating, for example, by favoring easily quantifiable phenomena and ignoring other, perhaps more important aspects.

I like to compare scientific research and its technical application with the exploration and opening up of unknown territories. As a first step explorers are sent out to crisscross uncharted terrain. Later a decision is made to build roads in the newly opened territory, an enterprise that profoundly changes the landscape. Moreover, many people will use these roads to get to their destination more quickly. Only a few will take the trouble to wander through still uncharted terrain. We become so accustomed to streets that we are not even aware that they obstruct our view of the landscape.

The natural sciences and technology find themselves in a similar position. Pure basic researchers are like those early explorers. They go into new territory, gather facts, and write everything down; and most of their records are destined to gather dust in libraries and archives. The next step involves technical exploration, the "road construc-tion." The builders of these roads rely on data gathered by the explorers. Finally a decision on the overall design of the road network is reached. The original explorers lose interest in the matter and turn their attention to other tasks. Soon the "roads" are completed; and industry is functioning and profiting from the early exploration.

There would be nothing wrong with this process if it were free of problems. Alas it is not. There is always the danger that with the passage of time one will overestimate the importance of existing roads. Finally one becomes so accustomed to the existing road network that its original purpose—to make the landscape accessible—is forgotten. One begins to think in terms of "roads," of one-dimen-sional causal chains, forgetting completely that at one time the roads were merely a means toward an end. They are nothing but one-dimensional structures for the explo-ration of a multidimensional reality.

Scientists are explorers. A scientist or technician who wants to succeed cannot be a one-sided rationalist, a "mere road builder," who accepts only that part of reality which is measurable. Particularly in the "birth phase" of scientific ideas, the unconscious, the ability to sense rela-tionships intuitively, plays a crucial part. The difference between a natural scientist and a composer or poet is less profound than is generally assumed. An acquaintance of mine, a professor of theoretical physics at a leading American university, proposed this thesis: a scientist needs twice the imagination of a poet or a composer. Why? Because, he said, a poet or musician creates some-thing new by himself without paying the slightest atten-tion to external conditions. A scientist also creates some-thing new, but at the same time he must take into account existing, external conditions laid down by nature.

262

It has become fashionable to indict scientists because of their allegedly one-sided emphasis on the investigation of rationally comprehensible phenomena. Scientists, engineers, and technicians are being held responsible for the problems of modern civilization. What is forgotten is that scientific discoveries are above all the fruits of human creativity, which ultimately rests on irrational impulses that cannot be measured and cannot be put into categories of causal relationships.

But one thing sets scientists apart from their critics. More than most others, they recognize the danger of one-sided, irrational thinking. They know how fragile is the rational foundation on which our civilization rests. For proof we need only look at Auschwitz, the Gulag, the mass murders of Palestinians or of American and French soldiers in Beirut.

Scientists confine themselves as far as possible to what can be encompassed by logic and to measurable, testable phenomena. The fact that they purposely limit their field of vision makes them more aware of the limits of their knowledge. And that also explains their awe and respect for the infinite complexity of nature.

All this leads me to the following conclusion. What we need today is a strategy that will bridge the comprehensive, intuitive grasp of nature and the purposeful application of science and technology. We need the roads, but we also need a great deal of landscape. We cannot afford to cover the entire landscape with roads.

As simple as this conclusion may seem, reaching individual decisions is a more difficult matter. Should we build more atomic power plants for producing energy and assume the risks connected with this? Or should we do without atomic power plants and produce energy by burning coal, thereby risking the chemical pollution of the atmosphere, with all its catastrophic consequences? Or should we do without certain fields of technology altogether? It is impossible to come up with a clear answer, for these are questions that have no unequivocal answers. Basically these are questions, like those of quantum theory, for which no clear answers exist. The questions Where is an electron located at this moment? and What is its velocity? have no clear answers; only probabilities can be given. However, we cease to deal in probabilities the moment we try to find the velocity of an electron through experimentation. Thus in the experiment we can find a velocity of 1000 kilometers/second—a clear answer. But then it turns out that this clear answer was arrived at through an intrusion into the system—through measuring the electron's velocity with the help of an instrument. We have influenced the system—in this case the physical system of the electron—from the outside. Our decision to determine its velocity has created a new situation—bearing out Niels Bohr's maxim that we are both spectators and participants in the phenomena of nature.

The principle of complementarity in its general application basically means nothing more than recognizing the contradictions in nature and society and resolving them through our actions. I mentioned the necessity of developing a strategy that would enable us to mediate between the exploration of nature by science and the intuitive grasp of the unity of the cosmos. Physics, through quantum theory, succeeded in developing a language that mediates between seemingly contradictory descriptions of the electron (that is, the reciprocal uncertainties of place and velocity). Will we succeed in finding an analogous strategy for overcoming the destructive antagonisms of modern industrial society?

The Universe of the Mind

T E N S of thousands of years ago mankind began its systematic exploration of the environment, and in the process changed it. The world we are living in today is marked by the changes then set in motion. With these changes a new world came into being, the world of symbols and language.

The philosopher Karl Popper proposed the division of the world into three parts. The first part (World one) is identified with the physical world, the real universe that exists independent of man. This part encompasses electrons, quarks, and all the systems based on them. The second part (World two) is the world of subjective feelings, of the unconscious, and of psychic properties. The third (World three) encompasses the entire world of conscious thought.

Like all classifications, Popper's is to some extent arbitrary. Who would venture to define precisely the line dividing World two from World three? Nonetheless, his approach is useful. I do not propose to go as far as Popper. For my purpose the division of the world into two parts is adequate. The first part is identical with Popper's World one, which I will call the physical universe, and the second embraces Popper's two remaining worlds. I should like to call this world of psychic conditions, of logical symbols and language, the universe of the mind, corresponding roughly to Plato's world of ideas. In contrast to the physical universe, this spiritual universe is the product of mankind. The utilitarian inventions of our conscious and unconscious mind are its elements, its symbols. To describe observed phenomena we invent concepts and symbols. Human thought consists in the meaningful combination of these symbols.

This suggested bipartition is analogous to the division used in computer science. The physical universe as a concrete system corresponds to the computer itself, to the hardware. The universe of the mind is analogous to the computer programs, the software. Of course, there is a significant difference between the real world and the world of the computer. A computer can function properly only if the program fed into it is free of errors. Everything must be clearly defined, without the slightest uncertainty.

The universe of the mind, on the other hand, is a living world, a world in flux. Only rarely are the symbols it encompasses clearly defined. Its concepts and symbols are full of uncertainties.

In a manner of speaking the universe of the mind corresponds to a computer program that is undergoing constant change. A computer whose program is constantly being changed could not function. The crucial difference between the intellectual activity of human beings and the activity of a computer lies precisely in the fact that the human mind does not collapse when new knowledge is acquired and new relationships evolve. Our intellectual universe is far more flexible than a rigidly constructed computer program.

Another essential difference appears when we look at the feedback from the universe to the physical universe. A computer program is not able to change the hardware. Human beings, however, are constantly changing their environment. There is constant interaction between the physical and mental universe.

In studying the phenomena of the physical universe we discover laws: the laws of nature. In days past, particularly toward the end of the last century, the laws of nature were believed to be more than merely the elements of the universe of the mind. They were seen as the essential properties of the physical universe, the skeletal structure of the world. The processes of the physical universe were seen as the interaction of the various laws of nature that interlocked like the wheels of a clockwork and determined the dynamics of the world.

Today this notion of the role of the laws of nature is undergoing radical revision. The laws of nature themselves are not elements of the physical world but of the universe of the mind. As such they are inventions of the human mind, even if inventions corresponding to essential aspects of the physical universe. But—and this is important—natural laws are not the skeleton of the physical world keeping it together; the universe is not in need of a skeleton. This recognition deprives the laws of nature, and with them the natural sciences, of that aura of the absolute that surrounded them until recently.

The universe of the mind—the world of intellect and ideas—is the most remarkable creation of mankind. Moreover, it is part of the eighth epoch of the universe. As mentioned earlier, only in the cold world of the eighth epoch can macroscopic structures and thus sentient beings exist, and thus only then can ideas and symbols exist.

At some time in the remote future this epoch will come to an end, and with it the world of ideas. The universe of the mind will disintegrate into nothing. No one will survive to tell of the end of the eighth epoch.

To describe their environment human beings invented concepts corresponding to specific aspects of reality. But often these concepts derived from views that had crystallized in the course of the evolutionary process. Our notion of space is a typical example of this conceptual approach. Intuitively all of us sense what the space surrounding us is all about. A straight line obviously must be the shortest distance between two points in space.

After centuries of physical and mathematical research we found out that our intuitive concepts of space and time do not necessarily conform to reality. The space around us holds a great deal more structure than was generally believed. Thus, gravitational fields (for example, the gravitational field of the earth) are able to change the structure of space. It turns out that a straight line may not after all be the shortest distance between two points; instead, a curved line, often with only a very small curvature, is. Also, it is not easy for us to imagine a three-dimensional space that is finite but without limits. Yet such spaces exist, at least in the abstract, conceptual world of mathematics, and it is altogether possible that our universe is just such a vast finite space.

We tend to believe that a finite space, however vast, must be contained in something else. Suppose our universe is finite. A number of questions immediately come to mind: What lies beyond it? Is there a boundary, a solid wall, dividing our universe from its surroundings? These questions have no answer; they are meaningless. A finite but boundless universe is a self-enclosed entity. There is no "outside."

Yet we continue to ask questions about the "outside" because we are used to finite volumes, for example, spheres that have an inside and an outside. The concepts and ways of seeing that we have developed for describing our environment are related to parts of nature, to systems embedded in larger systems. But the universe is not embedded in another system. It exists by itself and is not subject to our naïve way of looking at things. The concepts acquired in our humdrum lives cannot encompass the universe as a whole.

A similar problem crops up in connection with the finiteness of time. Our universe came into being about 20 billion years ago. The question that immediately comes to mind concerns what had been before, what had existed before the Big Bang. This question, too, is meaningless, because there can be no time outside the universe. Before the Big Bang there was nothing, neither time nor space nor matter.

Most people will find this answer unsatisfactory. Our sense of time tells us that at any given time it is possible to tell what had preceded it. We see time as something that flows along evenly, untouched by outside forces, like a river that cannot be stopped. We acquired this sense of time in the course of our evolution because in our day-to-day existence, time indeed behaves like a gently flowing stream. It seems as though nothing and no one is able to affect the passage of time. Processes on earth have as little effect on the architecture of space and time as a gentle breeze on the steel structure of a skyscraper.

A tall building cannot be toppled by a breeze, whereas a typhoon can pose a danger. Today we know of physical phenomena and processes that, like typhoons, are able to alter the structure of space and time irrevocably. Black Holes are such phenomena, and so is the Big Bang. The processes that took place in the cosmos soon after the Big Bang differ radically from those we encounter in our daily lives. They force space and time into a straitjacket. Space, time, and matter were closely intertwined in the first microseconds of the universe. Their constant interplay can be described and retraced within the framework of our cosmological theories. The result of this extrapolation is familiar to us: space, time, and matter have not been in existence for an infinite time. The product of the Big Bang, they came into being about 20 billion years ago.

These observations lead us to the following deduction. Although our concepts were gained through experience, natural science allows us to indulge in far-reaching extrapolations that can have a bearing on the entire cosmic scene. No rational explanation for the success of this method exists. It is and remains a miracle that the universe of the mind created by human beings enables us to offer testimony about extreme situations in the cosmos, such as the epochs following the Big Bang.

The universe of the mind is the product of reason. It encompasses all the knowledge accumulated by humanity in the course of its history. Today we see that this knowledge has jelled into a view of the world whose essential aspects were formed by the natural sciences. We can now say that we understand the most vital aspects of the dynamics of the microcosm and macrocosm. By changing our environment we are constantly creating new objects, new contexts, new ideas. It is a process that will go on as long as critical minds are at work. Consequently the universe of the mind is constantly expanding and re-forming. New dimensions of reality are opening up. Human creativity knows no bounds. Therein lies our great hope for the future as well as its great dangers. No one today can claim to command a complete view of accumulated human knowledge. The universe of the mind increasingly is being splintered into discrete sectors. It takes an expert to have an overview of its various, ever more specialized domains. Should we stand by helplessly to watch this splintering process?

We are being inundated by information that no one is able to integrate into a coherent overview of nature and the cosmos. This trend is sure to accelerate, particularly with the growing reliance on computers and microelectronics. The constant expansion of our knowledge combined with the increasing human intrusion into nature must not become a purpose in itself, an activity that ultimately loses contact with the needs of human beings.

The expansion of the universe of the mind, unlike the expansion of the physical universe, can be controlled. I believe the time for the exercise of such control has come. Untrammeled scientific research for the sake of research, the amassing of fresh knowledge for the sake of knowledge, is no longer desirable. Because the store of potentially attainable knowledge has no limits we must begin to set limits. We must acknowledge the fact that the scientific picture of the world can never be complete. It will remain forever incomplete. The universe of the mind, like music, has been created by us. And so it is up to us to keep it within bounds.

This is not the place to discuss possible ways of solving the dilemma of the splintered universe of the mind. But I do wish to stress one aspect I deem important. All scientists, engineers, and technicians must begin to show greater concern for incorporating their insights into the broad framework of our knowledge of nature. Except in special cases we will no longer be able to enjoy the luxury of research for its own sake. Scientist-specialists who cannot convince the general public of the importance of their work do not deserve either public or private support of that work.

We must also restructure the educational programs of our public schools. The greater the splintering into special areas of study, the greater the need to give students an overview of nature and the cosmos, rather than the sort of piecemeal information that only discourages them and ultimately does more harm than good.

In the past nature was thought to be the repository of all knowledge, and, so we believed, all that was needed was to accumulate as much of that knowledge as possible in the shortest possible time. Natural scientists, it was assumed, were a species of berry pickers out to gather as many berries as they could. The time has come to relinquish such naïve notions, to recognize the fallacy of that idea. Natural science and technology, like music and painting, are human inventions. Once we accept this, it will become possible to find a limit in the universe of the mind that enables us to set boundaries—though boundaries flexible enough for the sort of freedom essential to all successful research.

CHAPTER 17

God and the Absurd Universe

> [I]n a universe suddenly divested of illusions and lights, man feels an alien, a stranger. . . . Thus, convinced of the wholly human origin of all that is human, a blind man eager to see who knows that the night has no end, he is still on the go.
>
> —ALBERT CAMUS
> *The Myth of Sisyphus*

F O R centuries mankind has used reason and experimental research for its investigation of the structures of the microcosm and macrocosm. We have built powerful telescopes, sophisticated particle accelerators, costly measuring devices, and much more. In the process we learned an astonishing amount about the complex structure of the universe of the eighth epoch. But something has not been found in this search, neither in the depths of intergalactic space nor within the atoms and nuclear particles: the reason for our presence on earth and a way of arriving at ethical values and norms through the scientific method. Albert Einstein commented on this problem as follows:

The awareness of truth is glorious, but as a guide it is so powerless that it cannot even substantiate the justification and value of our pursuit of truth. Here we are simply faced with the limits of the rational understanding of our existence.*

The question about the meaning of life cannot be answered without taking mankind and its place in the unity of the universe into consideration. Every phenomenon, every object in the cold eighth epoch of the cosmos, represents an individual structure; yet it is also part of the whole. Human beings are no exception. That is why absolute freedom does not exist, nor absolute truth. If we ignore this only intuitively felt connection with the unity of the cosmos and rely solely on an objective, rational conception of the world, our search for meaning becomes futile.

We do not know why the earth exists. We do not know why the universe embarked on the adventure of the eighth epoch instead of putting an end to its existence earlier. We do not even know the reason for the Big Bang, nor do we know whether there was a reason at all. Questions like these and about what happened before the Big Bang are pointless. Looked at from the vantage point of objective knowledge, the universe makes no sense: it is absurd. No one saw this more clearly than Camus:

[I]n a universe suddenly divested of illusions and lights, man feels an alien, a stranger. His exile is without remedy since he is deprived of the memory of a lost home or the hope of a

*Albert Einstein, *Out of My Later Years* (Westport, Conn.: Greenwood Press, 1970), p. 22.

promised land. This divorce between man and his life, the actor and his setting, is properly the feeling of absurdity.*

The more deeply we penetrate into the structures of the cosmos, the more abstract, mysterious, absurd seems its architecture. The world familiar to us since childhood is gradually slipping away; new, previously unknown shapes begin to emerge. The world of galaxies and quarks is alien and can be understood only abstractly. In its cold and simple beauty it is absurd, without sense or purpose. Here are Camus's thoughts on this:

> If man realized that the universe like him can love and suffer, he would be reconciled. If thought discovered in the shimmering mirrors of phenomena eternal relations capable of summing them up and summing themselves up in a single principle, then would be seen an intellectual joy of which the myth of the blessed would be but a ridiculous imitation.†

The idea of God sprang up in answer to the question about the meaning of life. God is the cause of all being and order in the world. He can, it is believed, mediate the deeper meaning and is the source of eternal life after death.

Apparently the question about the meaning of life began to preoccupy human beings at an early stage in their development. In the search for an answer they developed myths and religions. Jacques Monod maintains that merely posing the question of the meaning of life fulfills an inborn need: "That this imperious need develops spontaneously, that it is inborn, inscribed somewhere in the genetic code, strikes me as beyond doubt."‡

God is not only that first cause lending meaning to our

*Albert Camus, *The Myth of Sisyphus* (New York: Vintage Books, 1955), p. 5.
†*Ibid.*

‡Jacques Monod, *Chance and Necessity*, trans. Austryn Wainhouse (New York: Knopf, 1971), p. 167.

existence. He is also the God concerned with our everyday worries, whose reward is desired and whose punishment feared.

What about the existence of God in our world of science? Is there room for God in a world that seemingly can dispense with His intervention in the processes of the universe? And what about God if the universe of the mind ultimately also disappears and nothing is left but an increasingly hot or cold gas of photons, neutrinos, and other particles? Are religion and science an irreconcilable contradiction?

I do not believe that such a contradiction exists nor ever existed. The scientific method of exploration concerns the creation of appropriate concepts to describe natural phenomena and the establishment of connections between these phenomena. We feel our way along this "road of ideas," along chains of causality. We open up reality by attempting to construct a logical "road network" free of contradictions, one which we then clap on to reality. I have cautioned the reader against confusing the network of scientific concepts—our picture of reality —with the real world. Many of the mistakes and much of the tragically flawed reasoning of our time are based on just this fallacy. The world is more than merely such a network of roads. It is a continuum that cannot be completely encompassed by even the most finely spun net of one-dimensional "roads of causality." Only through the intuitive feeling of the unity of the cosmos will we be able to glimpse the continuum outside the "roads." This intuitive sense for the unity of the universe in my opinion warrants the designation *religiosity*.

When I speak of religiosity I have in mind primarily but not exclusively the cosmic belief Spinoza and Einstein spoke of. This religiosity affirms the unity of being and sees the individual as part of the unity of nature. Such a religiosity has no room for a personal god who rewards and punishes. It is part of our universe of the mind, a product of the human spirit like music and art, a product of our culture.

The Complementarity of Science and Religion

I do not think of science and religion as adversaries. Quite the contrary. They are two complementary sides of our view of the cosmos. They are interdependent. In Einstein's words: "Science without religion is lame, religion without science, blind."*

When I speak of the complementarity of science and religion I do not mean to imply that science has made religion obsolete. Naïve reliance on progress is a thing of the past. Gone are the days when we thought that accumulating knowledge and harnessing nature will bring us more profound insight into the meaning of life and lead automatically to a better world. Nothing is worse than the sort of vulgar belief that sees the meaning of human existence *only* in the acquisition of material values, the permanent expansion of technical possibilities, and an ever deeper involvement of science in all aspects of life.

We have seen that the universe forms a unity, that it has a history, and that it will die. I must confess that to a large extent the appeal the cosmos holds for me lies in these aspects. Since Einstein and Hubble and the discoveries of particle physicists, the universe has become less alien to us.

*Einstein, *My Later Years,* p. 26.

We know that the evolution of the cosmos during the past 20 billion years has been a continuous process marked by the creation of new structures as well as the death of old ones. Nothing lasts forever. Death, including our own death, is an intrinsic part of the processes taking place in the eighth epoch.

Religiosity is well able to lend meaning to life and provide a structure for ethical values and norms. What Camus and Monod failed to find in the cold, rational world of quarks and galaxies can be found in the necessity for human beings to see themselves as part of the cosmos. We are neither rulers nor slaves of the world surrounding us, but are embedded in the inexhaustible continuum of possibilities and relations the eighth epoch holds out to us. We are the product of that epoch and carry within us the traces of its varied history since the Big Bang. At the same time our actions allow us to shape the course of history. All our actions are unique in the cosmos, just as every second is unique. Therein lies the meaning: in living; in the experience of the cosmic unity that transcends the individual; in the self-consciousness of the acting individual aware of his or her limits and from this awareness deriving the confidence essential to the acceptance of life.

God is therefore also the unity of the universe in its eighth epoch, which reveals itself to us in diversity. He, like ourselves, is a part of the eighth epoch. Thus if we understand *eternal* in its temporal-physical sense, there is no eternal God, just as there is no eternal life.

Starting from this vantage point, when I look at the religions of the world I am struck by the similarity between this conception and the ideas of the major Eastern religions, of Hinduism, Buddhism, and Taoism. Hinduism has thousands of gods, all of them ultimately different embodiments of one and the same divine reality expressed in the unity of the cosmos. The most famous of these gods, Shiva, is the god of creation and annihilation —the symbol of the dynamics of life processes.

Buddhism's road to self-realization, the way to the achievement of a divine state of consciousness, of Nirvana, is, like the evolution of the universe, eightfold. Detachment from the individual, the experience of wholeness, are essential features of Buddhism.

It is illuminating to find that the Taoism of ancient China preempted significant insights of modern science. Thus Taoism teaches that reality is constant change marked by the emergence of stable structures. Taoism sees the dynamics of the world as a constant interchange between conflicting positions, between yin and yang, as expressed by this aphorism in Lao Tzu's *Way:*

> To remain whole, be twisted!
> To become straight, let yourself be bent.
> To become full, be hollow.
> Be tattered, that you may be renewed.*

All Eastern religions have in common an awareness of the unity of the cosmos and its significance. They recognize that the division, or "dissection," of the cosmos is a human construct, not an inherent property of the universe. Hinduism even goes so far as to see the human trait of differentiating, of categorizing, as a sickness to be cured through meditation.

A comparison between Eastern religions and Christianity points up an interesting difference. The suffering indi-

---

*Arthur Waley, *The Way and its Power* (Grove Press, n.d.), p. 171. Reprinted with permission of Grove Press, Inc.

vidual seeking forgiveness, one of the leitmotifs of Christianity, plays only a minor role in Eastern religions. The God of Christianity has human features; He concerns Himself with the suffering and sorrows of the individual. Therein lie both the power and success as well as the weakness of Christianity. I believe the Eastern religions can be brought into harmony with the insights of modern science; I see problems with Christianity. The God of Spinoza and Einstein, the God of unity in diversity, is quite compatible with the Eastern religious concept of God, but not so readily with the God of Christianity.

For almost four hundred years, since the Vatican's trial of Galileo, science and the Church have been at odds. To this day the Catholic Church has not clearly stated that its verdict on Galileo, although relatively mild compared to others of that time, was a mistake that has been a centuries'-long obstacle to a sensible dialogue between religion and science. Today's world is also the world of science and technology. It is only a question of time before everyone, not only those involved in research, will become concerned about the relationship of science and religion. The time of reciprocal exclusion, or division into discrete competencies, is past.

The world of belief and the world of science are complementary, interdependent worlds. To draw a sharp dividing line between them, as many theologians have done in the past and still do today, is absurd.

I see no direct contradiction between the Christian churches and science. But problems of interpretation of religious symbols do exist. These symbols do not represent immutable, everlasting values. Like the concepts of science, religious symbols and rites are an expression of our desire to overcome the inadequacy of language in our quest for the sublime.

The historical nature of Christianity's beliefs cannot be ignored. The Bible, too, is primarily a historical document subject to reinterpretation. Given the discoveries of Darwin and Einstein, of Hubble and Gamow, of molecular biologists, astrophysicists, and particle physicists, such interpretation is badly needed. What we lack are theologians willing to undertake the task.

Voices can be heard accusing science and technology of having brought civilization to the edge of an abyss. I cannot join in this chorus. True—the abyss is there. The coming decades will show whether humanity is capable of defusing the threatening danger of self-annihilation by atomic, chemical, and biological weapons or by the pollution of the environment. Yet this abyss was not opened up by science and technology but by the human inability to solve conflicts. I am convinced that the insights science, and thus philosophy, have brought us are bound to play a significant part in resolving the conflicts between different social systems and different social groups. The knowledge gained through science is the tie binding all of us.

We are the first in time to know where we stand in the universe. Modern science has given us the humility appropriate to sentient beings on a small planet near the galactic Virgo cluster. My knowledge of the cosmos has not only made me humble but also has given me a measure of composure and pride. We know that our position in the cosmos is not unique. No longer caught up in the myths of the past, we now know where we stand and what we are. Each one of us is unique. Each one of us plays a role, however minor it may seem. There are no extras on our stage.

You might ask the point of all this if some day in the remote future we will all be gone, the only trace of all our striving and ambitions a thin gas of light particles with no sense of the past. I cannot agree. We are here. The planet glowing with a faint blue light amid the darkness of the cosmos is our home. This insignificant star that supplies us with vital energy is our star. This galaxy near the Virgo cluster is our galaxy. And the time in which we are living, 20 billion years after the beginning, is our time.

In the future we will have to focus more strongly on the earth. No one can exist in the universe detached from the earth, detached from the 20 billion years of evolution. All of us are enmeshed in the complex network of relationships and dependencies of the eighth epoch without which there could be no life, no awareness, no human spirit, no God. Absolute freedom does not exist, neither in the intellectual nor in the moral sphere.

The universe is more than merely an accretion of electrons, quarks, and galaxies, more than space and time. The complex, interrelated world of the earth which created us is part of it. It is our duty not only to ourselves to preserve this world. The universe itself imposes this task on us.

Appendix

Useful Powers of Ten

I N science we frequently deal with objects that, measured by ordinary yardsticks such as human height, are immense or minute (for example, the volume of a galaxy or an atom). If the size of these objects were given in our customary units of measurement we would get either enormous or diminutive figures. It is therefore convenient and useful to define such quantities in figures ranging from 1 to 10 multiplied by the appropriate power of 10. Thus, for example, we write the figure 850 as 8.5×10^2, or 0.0031 as 3.1×10^{-3} (10^{-1} stands for 1/10, 10^{-2} for 1/100, 10^{-3} for 1/1000, and so on). One of the advantages of this method is that it makes multiplication easy, for in multiplying two powers of ten all we need do is add their exponents. For example, $10^5 \times 10^2 = 10^7$, $10^5 \times 10^{-3} = 10^2$. Powers of ten simplify estimates of orders of magnitude. Both the imaginable and the unimaginable can be described with their help. For example, the mass of an adult on the average amounts to 7.5×10^1 kilograms, that of the earth to 6×10^{24} kilograms, and that of the sun to 2×10^{30} kilograms. The number of quarks in the universe is estimated at 10^{80}. A human being is composed of about 1.5×10^{29} quarks.

Glossary

Alpha particle: The nucleus of the helium atom. It consists of two protons and two neutrons. Alpha particles are radiated by some radioactive substances (alpha radiation). Frequently abbreviated as α particle.

Andromeda nebula: Our closest galactic neighbor, composed of about 300 billion stars and located about 20 million light-years away from the earth.

Antiparticle: A particle with the same mass and spin as the particle in question but with opposite electric charge, baryon number, lepton number, and so on. For every particle, there is a corresponding antiparticle. Certain purely neutral particles, such as photons and π^0 mesons, are their own antiparticles. The antineutrino is the antiparticle of the neutrino, the antiproton is the antiparticle of the proton, and so on. Antimatter consists of antiprotons, antineutrons, and antielectrons (normally called positrons).

Asymptotic freedom: The diminution of forces between quarks at short distances. A phenomenon of chromodynamics.

Baryons: A class of strongly interacting particles, including neutrons and the unstable hadrons known as hyperons. Baryon number is the total number of baryons present in a system minus the total number of antibaryons.

Beta decay: The decay of the neutron into a proton, an electron, and an antineutrino. This decay is a consequence of the weak interaction. Frequently abbreviated as B decay. The weak decay of a nucleus is also called B decay.

Boson: A name for all particles that have integral spin. Examples of bosons are the π meson (spin 0), the photon (spin 1), and the W boson (spin 1).

Bubble chamber: An instrument for detecting particles. It consists of a vessel filled with a fluid heated nearly to the boiling point. When the pressure on the fluid is diminished suddenly, the fluid becomes overheated. At this moment electrically charged particles are shot through the bubble chamber. The fluid begins to boil along the paths of the particles, developing steam bubbles that can be photographed.

Chromodynamics: A theory of the interaction between quarks and gluons. Physicists believe that quantum chromodynamics (QCD) is the correct theory of the strong interactions.

Cluster: An accumulation of numerous galaxies; they can number in the thousands. Both the Virgo and Coma clusters contain thousands of galaxies.

Glossary

Cosmologic principle: The hypothesis relating to the isotropy and homogeneity of the universe. This principle states that except for some local differences the structure of the cosmos appears identical to all observers in the cosmos.

Deuteron: The nucleus of deuterium. It consists of one proton and one neutron.

Electron: The lightest massive elementary particle. All chemical properties of atoms and molecules are determined by the electric interactions of electrons with each other and with the atomic nuclei.

Electron volt: A unit of energy equal to the energy acquired by one electron in passing through a voltage difference of 1 Volt. Equal to 1.602×10^{-19} watt· second. Often the units MeV ($= 10^6$ eV) and GeV ($= 10^9$ eV) are used.

Energy: A physical property which in conjunction with impulse characterizes the motion of a particle. According to Einstein's theory of relativity the energy of a resting particle is proportional to its mass.

Fermion: A genetic term for all particles whose spin is $\frac{1}{2}$.

Fine structure constant: The fundamental numerical constant of atomic physics and quantum electrodynamics, defined as the square of the charge of the electron divided by the product of Planck's constant and the speed of light. Denoted by α and equal to 1/137.-036.

Glossary

Friedman's model: The space-time model of the universe based on general relativity developed by the Russian mathematician Alexander Friedman.

Galaxy: A large cluster comprising up to 1000 billion stars held together by gravitation. Galaxies appear in a variety of shapes—ellipses, spirals, beams—as well as irregular forms.

Gauge theories: A class of field theories currently under intense study as possible theories of the weak, electromagnetic, and strong interactions. Such theories are invariant under a symmetry transformation, whose effect varies from point to point in space-time.

General relativity: The theory of gravitation developed by Albert Einstein in the decade 1906–1916. As formulated by Einstein, the essential idea of general relativity is that gravitation is an effect of the curvature of the space-time continuum.

Glueball: A neutral meson consisting solely of gluons.

Gluons: Electrically neutral objects that mediate the interaction between quarks within the framework of chromodynamics. Gluons have spin 1.

Gravitation: The phenomenon of the mutual attraction of massive objects. According to the theory of general relativity, gravitation is a consequence of the change wrought in the structure of the space-time continuum by the presence of matter.

Group: A mathematical system of different (sometimes

infinitely many) elements, with very definite rules. For example, the multiplication of two elements within the group is a well-defined operation. Group theory plays a very important role in physics. For example, it allows us to describe the symmetry of particles in a very simple manner.

Hadron: A particle that participates in the strong interaction. Hadrons are divided into baryons (such as the neutron and proton), which consist of three quarks each and obey the Pauli exclusion principle, and mesons, which do not obey this principle. Baryons have nonintegral spin (1/2, 3/2, . . .); mesons have integral spin (0, 1, 2, . . .).

Helium: The second most numerous chemical element. Helium appears in nature as ^{4}He, with an atomic nucleus composed of two protons and two neutrons, and as ^{3}He, with an atomic nucleus consisting of two protons and one neutron.

Hubble's Law: The proportionate relationship, first observed by Edwin Hubble, between the velocity with which remote galaxies are moving away from our galaxy and their distance. The relationship between velocity and distance is known as Hubble's parameter.

Interaction: Two objects are said to be interacting when they exert influence on each other (for example, through forces).

Jet: A system of particles produced during particle reactions at high energies. The jets are interpreted as fragments of elementary objects such as quarks and gluons.

Kelvin scale: A temperature scale whose zero point coincides with the absolute zero of temperature. At an atmospheric pressure of one atmosphere the melting point of ice is 273.15°K.

Leptons: A class of particles that do not participate in the strong interactions, including the electron, muon, and neutrino. Lepton number is the total number of leptons present in a system minus the total number of antileptons. Leptons have spin 1/2.

Light-year: The distance covered by light in one year: 9.46×10^{12} kilometers.

Maxwell's equations: A group of equations that describe the dynamics of electromagnetic fields, derived by James Clerk Maxwell in the nineteenth century.

Mesons: A class of strongly interacting particles, including the π mesons (also called pions), K mesons, ρ mesons, and so on, with zero baryon number.

Neutrino: A massless or nearly massless electrically neutral particle, denoted by ν. It participates only in weak and gravitational interactions and comes in at least three varieties, known as the electron-neutrino (ν_e), the muon-neutrino (ν_μ), and the tau-neutrino (ν_τ).

Neutron: The uncharged particle found along with protons in ordinary atomic nuclei; denoted by n.

Nuclear fusion: The fusion of light atomic nuclei into heavier nuclei in which large quantities of energy are released.

Nucleon: Generic term for protons and neutrons.

Photon: In the quantum theory of radiation, the particle associated with a light wave or, more generally, an electromagnetic wave; denoted by γ.

π Meson: The hadron of the lowest mass. Comes in three varieties, a positively charged particle (π^+), its negatively charged antiparticle (π^-), and a slightly lighter neutral particle (π^0). Sometimes called pion, the scientific notation is π meson.

Planck's constant: The fundamental constant of quantum mechanics; denoted by h. Planck's constant was first introduced in 1900, in Planck's theory of blackbody radiation. It then appeared in Einstein's 1905 theory of photons: the energy of a photon is Planck's constant times the speed of light divided by the wavelength. Today the constant h is more frequently used, defined as Planck's constant divided by 2π. The numerical value of h is 6.6×10^{-34} watt·second^2.

Planck's elementary length: In quantum theory the bottom limit of the measurement of length determined by gravitational interaction. For distances smaller than Planck's elementary length our normal concepts of space and time become meaningless. Planck's elementary length is numerically given as 4×10^{-33} centimeters.

Plasma: A hot gas composed of interacting particles.

Positron: The positively charged antiparticle of the electron; denoted by e^+.

Proton: The positively charged particle found along with neutrons in ordinary atomic nuclei; denoted by p. The nucleus of hydrogen consists of one proton.

Quantum electrodynamics: The quantum theory of electromagnetic phenomena (QED).

Quarks: The constituents of nucleons. To date, the existence of five different types of quarks has been established (u, d, s, c, b). It is assumed that quarks cannot be produced as isolated particles. Stable matter is composed of two quarks, u and d.

Quasars: Astronomic objects displaying a substantial red shift. It is now generally assumed that quasars are the very bright nuclei of remote galaxies.

Red shift: The magnification of the wavelength of light emitted by a source moving away from the observer. By measuring the red shift of the light emitted by remote galaxies we can estimate the speed with which the galaxy is moving away.

Relativity, Theory of: The theory developed in the beginning of this century, primarily by Albert Einstein, which made us understand the dynamics of rapidly moving bodies. Within the framework of this theory, space and time are joined into a single unit. We tend to differentiate between the special and the general theory of relativity. The general theory of relativity incorporates gravitational interaction.

Speed of light: The fundamental constant of special relativity, equal to 299,729 kilometers/second; denoted by c. Any particles of zero mass, such as photons or neutrinos, travel at the speed of light. Massive particles approach the speed of light when their energy is very large relative to their rest energy.

Spin: A fundamental property of elementary particles that describes their state of rotation. According to the rules of quantum mechanics, spin can have only certain special values and is always either an integer or a half integer (1/2, 3/2, 5/2, . . .) multiplied by h.

Storage ring: A ring-shaped vacuum tube surrounded by a special magnet system serving the acceleration and storage of particle rays. In general two different types of particles are stored, for example electrons and positrons. The particles are allowed to collide inside appropriate detection instruments. During such collisions it is even possible to transform all the available kinetic energy of the particles into mass. That is why experiments in storage rings are energetically more favorable than the customary experiments in particle physics in which a moving particle is steered toward a target particle.

Superconductivity; Phenomenon appearing at very low temperatures in a series of electronic conductors, marked largely by the disappearance of electric resistance and the complete or partial displacement of an outer magnetic field from the conductor.

Supernova: A stellar explosion in which most of the

stellar matter is hurled out into interstellar space. This explosion releases as much energy as is radiated by the sun in billions of years. The last supernova in our galaxy was observed by Johannes Kepler in 1604.

Thermal equilibrium: A condition at which all particles possess the same mean energy. Every closed physical system eventually achieves thermal equilibrium.

Vacuum polarization: The change of the physical properties of space in the neighborhood of a particle which interacts with other particles. For example, the space around an electron is filled with virtual positrons that influence the distribution of the electric charge of the electron.

Index

absolute zero, 177
accelerators, 5, 110–11, 115–17, 126–29, 273; circular, 126, 154; linear, 126
action, 68
alpha beta gamma ($\alpha\beta\gamma$) theory, 218–219
alpha (α) particles, 50, 130, 131
Alpher, Ralph, 218
aluminum, atomic structure of, 73
Anderson, Carl, 109–10, 111–12
Andromeda constellation, 24, 31
Andromeda nebula, 24–26, 27, 29, 31, 33–35, 191; distance to, 201; distribution of stars in, 33–34; hydrogen clouds in, 34; Milky Way vs., 25, 27, 33, 35, 39; supernova explosion in, 31–32
antimatter, 108–27; collisions of matter with, 111, 112–13, 114–

119, 125; in evolution of universe, 225; galaxies of, 125–26; symmetry of, 125
antineutrinos, 119; composition of, 136; from proton decay, 169
antiprotons, 110–11, 114; collisions of, with protons, 114–19, 138; composition of, 136
antiquarks, 136, 138, 150, 152, 164, 181, 189, 190; annihilation of, with quarks, 226–27; in cosmic contraction, 250; in evolution of universe, 226; in thermal equilibrium, 188; from X-particle decay, 186
anti-X particles, 186–88, 189
Aristotle, 20
astronomy, 18–19; radio, 34, 41
astrophysics, 5, 8, 18–19
asymptotic force, 149
atomic nuclei, 50–52, 69; instabil-

Index

color (chromodynamic) force, 253, 257; distance and, 144–45, 149, 227; electric force vs., 143–44, 148–49; as manifestation of fundamental force, 163, 164; nuclear force resulting from, 157; strength of, 148–49; "white" product of, 146–48; X force vs., 164

Coma cluster, 194, 195–98

common sense, rejection of, 15

complementarity, principle of, 264

cosmic rays, 108–09, 166, 231–32; in cloud chambers, 109; in spark chambers, 108

cosmological principle, 198, 201–2, 224

cosmos, *see* universe

Cosmos (Sagan), 19n, 23

creation myths, 19–20

Creation of the Universe, The (Gamow), 219

Cronin, James W., 186

Crux constellation, 29

Curie, Marie, 173

Cygnus constellation, 29

Cygnus X-1, 44

Darwin, Charles, 251, 281

Davy, Humphry, 100

Democritus, 48–49, 55, 75, 258

DESY (German Research Center), 127

determinism: Newtonian physics characterized by, 77–78, 89; quantum mechanics vs., 78–79, 89

deuterium, 230

dinosaurs, evolution and extinction of, 236

Dirac, Paul A. M., 110

Dirac's equation, 110

distance, astronomic, 30–32, 192–93; time factor and, 31–32

DNA, structure of, 73–74, 257

d quarks, 134–36, 156, 186, 228, 253, 255

dwarf stars, 37

earth: evolution of life on, 235–36; formation of, 235

educational institutions: specialization fostered by, 16–17; understanding of science neglected by, 17–18

Einstein, Albert, 13, 16, 41–43, 57, 63, 237, 253, 276–77, 280; on electromagnetic waves, 105–6; on gravitation, 93–96, 97, 207–8; mass-energy conversions according to, 111; on meaning of life, 274; probability interpretation rejected by, 83–84

electric charge, 53–59, 98–99; absence of, in neutrinos, 121; absence of, in neutrons, 55–56, 166; of electrons, 53, 58, 159; gravitation vs., 99; of lepton-quark particles, 163; of leptons, 160–61; magnetism and, 99–100; of protons, 58, 159; of quarks, 135–36, 139–40, 156, 159, 160, 161; units of, 58, 135, 159

electric force: color force vs., 143–44, 148–49; distance and, 143–

Index

Index